INTERNATIONAL HARVESTER EXPERIMENTAL AND PROTOTYPE TRACTORS

Guy Fay

First published in 1997 by Motorbooks International Publishers & Wholesalers, 729 Prospect Avenue, PO Box 1, Osceola, WI 54020-0001 USA

© Guy Fay, 1997

All rights reserved. With the exception of quoting brief passages for the purposes of review no part of this publication may be reproduced without prior written permission from the Publisher

Motorbooks International is a certified trademark, registered with the United States Patent Office

The information in this book is true and complete to the best of our knowledge. All recommendations are made without any guarantee on the part of the author or Publisher, who also disclaim any liability incurred in connection with the use of this data or specific details

We recognize that some words, model names and designations, for example, mentioned herein are the property of the trademark holder. We use them for identification purposes only. This is not an official publication

Motorbooks International books are also available at discounts in bulk quantity for industrial or sales-promotional use. For details write to Special Sales Manager at the Publisher's address

Library of Congress Cataloging-in-Publication Data

Fay, Guy.
   International harvester experimental & prototype tractors/Guy Fay.
     p. cm.
   Includes index.
   ISBN 0-7603-0232-4 (pbk.: alk. paper)
   1. IHC tractors. 2. Experimental automobiles. I. Title.
TL233.6.I38F38   1997
629.225'2—dc21   96-49190

On the front cover: A color archival photograph of a Farmall Super M. The Super M pioneered a number of new features, including the torque amplifier and the independent power take-off. *State Historical Society of Wisconsin*

On the back cover: Top: The F-22 experimental tractor shows the evolution from the F-20 to the Farmall H. This tractor was never put into production. *State Historical Society of Wisconsin* **Bottom:** The T-65 TracTracTor experimental crawler photographed June 28, 1935. *State Historical Society of Wisconsin*

Printed in the United States of America

Contents

	Acknowledgments	4
Chapter 1	Reapers, Binders, and Engines	5
Chapter 2	Early Tractor Development	12
Chapter 3	Light Moguls and Titans	23
Chapter 4	Development of the McCormick-Deering Tractors	40
Chapter 5	Motor Cultivator and Farmall Development	54
Chapter 6	The Improved Power Program	75
Chapter 7	12 Series Tractors	84
Chapter 8	Letter Series Development	92
Chapter 9	After the War and Beyond	110
Chapter 10	Crawler Development	130
	Appendix	158
	Index	160

Acknowledgments

Like most large projects, this book was produced with a large amount of help and guidance. The list is long, and it includes Cindy Knight, the McCormick-International Harvester Archivist at the State Historical Society of Wisconsin, who manages the McCormick-International Harvester Archives. In addition, Nicolette Bromberg, Andy Krauschaar, Nez Zaragoita, and Scott Portman all helped with the images seen in this book.

Lee Klancher, my editor at Motorbooks, deserves thanks for getting this project off the ground as well as seeing it to completion, with both editing and general advice. Fellow tractor book authors Randy Leffingwell and Hans Halberstadt also helped me understand the process.

Case-IH also provided major support for this project through access to records and assistance in having them replicated, as well as access to employees. Dave Rodgers in Racine, and Steve Hiles and Rich Seraga at Case's Technical Center in Burr Ridge, Illinois all went out of their way to assist with this book.

Retired and ex-IH engineers and employees also assisted with this book. Al Allori, Ed Gaul, Gorden Herschman, and others assisted in telling the stories of the days when IH was great. In addition, current owners of IH experimental tractors, including Neil Stone and Wayne Hutton, as well as article author John McNaull, also assisted the author with descriptions of their machines.

On the editorial end, the author was assisted greatly in proofreading by Becky Soderblom, who edited the early drafts, and Mom, who edited the later manuscript. Proofreading for the tractor content was done by Jim Becker, Leroy Baumgardner, and Scott Saterlund, all of whom gave valuable assistance.

The Antique Tractor list on the Internet, administered by Spencer Yost, also was helpful in keeping me motivated as well as answering a few questions. The list is a truly interesting gathering of tractor lovers from many different nations.

And of course, my family, who in their various ways and forms all encouraged me throughout this process. Thanks, Grandma!

Chapter 1

Reapers, Binders, and Engines

The history of International Harvester (IH) starts more than 70 years before the company was formed, as does involvement with experimental machinery. In fact, the company roots were planted when Cyrus H. McCormick demonstrated a reaper before neighbors at Steele's Tavern, Virginia, site of his family's home, in 1831. History declared McCormick the "inventor" of the reaper. Like most of IH's history, there is controversy and debate over this seemingly simple fact. In later years, certain members of the McCormick family claimed that the reaper was primarily an invention of Cyrus' father, Robert. Despite the controversy, the younger McCormick was able to hit upon the happy combination of production and sales and eventually surpassed his competitors in the reaper business. He did so thanks to the eventual success of that early reaper, an experimental machine that eventually saw production.

Experimentation with technology did not stop with the reaper. Soon, self-tying binders would become more important. John Appleby of Mazomanie, Wisconsin, had invented an excellent twine binder and approached several manufacturers. Time after time, he was refused, either because of the newness of the invention (businessmen of the time were very conservative) or because they were afraid of the twine problem. Twine at that time was either expensive or fragile. Finally, Appleby had some success with a company in Beloit, Wisconsin, that made a few successful binders (this company eventually moved to Milwaukee and became the Milwaukee Harvester Company). People at the firm of Gammon & Deering then became interested in the idea. William Deering, one of the principles of the firm, operated a few Appleby binders with success in 1879 and he scheduled the manufacture of 3,000 binders for 1880.

Despite this ambitious schedule, Deering did not yet have good twine. The twine made then was not suitable for binder use. Deering's family had been in the textile business, and he himself had worked in the textile mills before moving west to Illinois. He untwisted a manila rope and tried using the strands, which convinced him manila was practical if a thin enough strand could be made. He approached several local rope makers without success. Eventually Deering had success in Philadelphia, where he was referred to Edwin H. Fitler. After

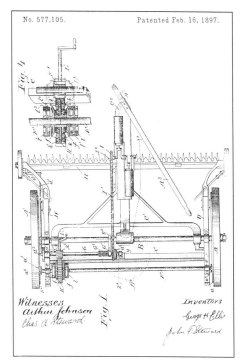

This is a top view of Ellis and Steward's first Automower patent. The machine is steered with a tiller (A3) by an operator sitting on the seat (A4, dotted lines). The two-cylinder opposed engine is clearly shown. The cutter bar is driven by a crank and rod on the left side of the machine. *U.S. Patent*

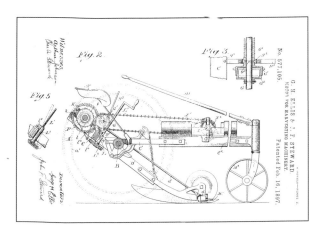

The side view of the first Deering Automower patent. The front end of the machine is truncated in the drawing. The Chain I takes power from the engine and brings it to the rear of the machine where it drives both the wheels and cutter bar. A novel machine, but probably very dusty for both engine and operator. *U.S. Patent*

first refusing Deering, Fitler agreed to the experiment on the promise of an order for several carloads of twine if successful. Fitler adjusted his rope machines, produced a few balls, and sent them to Texas, where John F. Steward was testing binders in the field.

The first ball of twine made by Fitler was successful, resulting in a telegram that read, "Manila Splendid." Deering was not only in the binder business but in the twine business as well. Deering went on to become McCormick's greatest competitor. McCormick itself began producing binders. The companies grew into great rivals. If one developed a machine, the other followed with a similar machine.

Engines

In the late 1880s, Deering had hired a young experimenter named George Ellis. He was hired as a twine manufacturing expert and over his lifetime received many patents in this area. Beyond twine, Ellis had mechanical skills and assisted with developing other machines, but he is best known for the internal combustion engine. Ellis started designing an engine in 1890 after the Otto patents on the basic gas engine had expired. He tried to interest William Deering in his first gasoline engine, which Ellis built in his basement just east of Deering Works, the huge factory that produced the Deering implements. Finished in 1891, this first engine had two parallel vertical cylinders and developed 6 horsepower. The engine had a 2 1/2-inch bore and a 6-inch stroke and weighed 22 pounds minus the two 24-pound flywheels. Cylinders were simple 1/8-inch Shelby steel tubing.

According to one history, Deering was "greatly impressed" with the engine and its steady operation, but wished the experimentation to remain secret (no doubt to avoid McCormick interest). Experimentation on the engine was transferred to yet another basement—that of the legendary John F. Steward. Steward had a home at Ravenswood, Illinois, where the new engine was placed on a New Ideal mower. The mower was demonstrated in October 1991 for Deering, Steward, and B. A. Kennedy, who eventually became a vice-president at IH. This was the first self-propelled farm implement built by any International Harvester predecessor company. Yet another first was built in Ravenswood: a horseless carriage propelled by the 6-horsepower engine.

Ellis began building a 16-horsepower engine that quickly began finding its way into other equipment, including a corn picker designed by the assistant superintendent of the Deering Harvester Company, Mr. Pitkin. During the December 1892 test, the horses pulling the picker were frightened by the auxiliary engine powering the picker mechanism and took off for parts unknown. No one was willing to loan the Deering folks a new team, so the experimenting was done for the day!

At some time in 1893 the 16-horsepower engine

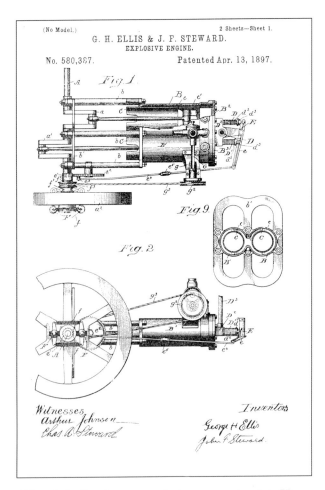

Although the first Automower had an opposed engine, Ellis and Steward experimented with a side-by-side two-cylinder engine as well. The valve motion, of course, was very strange, but Ellis and Steward's experiments led to the first IH horizontal stationary engines a few years later. *U.S. Patent*

Deering had a chrome- or silver-plated Automower built as part of its retrospective display of harvester development. The display was put together by J. F. Steward, one of the Automower's designers, and was meant to disprove Cyrus McCormick's claim to have invented the reaper. After the World's Fair was over, a French museum asked for and received the entire Deering display, presumably including the Automower. Another Automower was built for the actual field competition. *State Historical Society of Wisconsin*

was placed on a highly modified New Ideal mower. Over the next 9 or 10 years, a few of these were built for experimental and demonstration purposes. Although they evolved over the years, the engine/mower combinations were always known as "Deering Automowers." The 6-horsepower automobile was operated until the fall of 1895, when Ellis built a 12-horsepower engine and a new chassis to go with it. In January 1896, both the automobile and Ellis were shipped south to St. Augustine, Florida, where Ellis drove it for the Deering family. Although Deering refused to place the engine into production, he did encourage Ellis to keep experimenting with gasoline engines. He was especially interested in engines for farm use. Supposedly he was also interested in producing automobiles, but that would have to wait for some time in the future. "For the present they couldn't cover too many eggs," said Deering.

As far as it is known, the Deering Automower was never produced for sale, but was intended to be a demonstration machine along the lines of the "Dream Cars" that Detroit put out decades later. Deering Automowers appeared in the 1900 World's Fair, where they competed and lost to a McCormick Automower. George Ellis demonstrated a Deering Automower at the St. Louis Exposition in 1903, after which the Automower faded into obscurity.

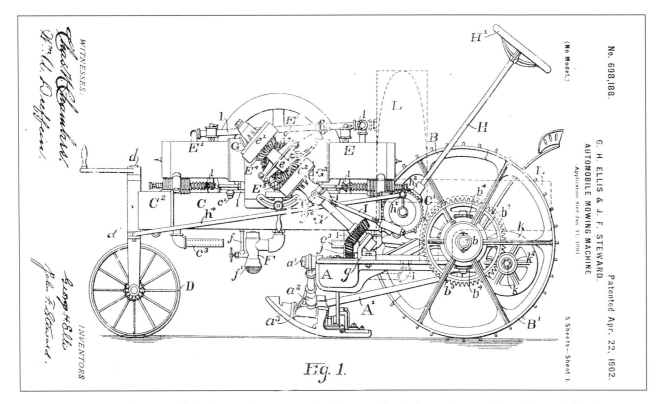

This patent shows the final form of the Deering Automower. A twin opposed-cylinder engine was still used, but with heavier construction. Shown is the PTO which consisted of shaft G coming off the engine. The gearing (I and J) transfer the power to a shaft that ends at A1 and extends back to the axle gearing. Power is taken from that location and transferred to shaft A1, which drives the mower shaft. *U.S. Patent*

John F. Steward

John F. Steward had an incredible history in the harvester business, predating the binder and ending with harvesters powered by gasoline engines. When not participating in the harvester business, he was exploring the Wild West (including a historical expedition down the Colorado River) or writing about the early history of Illinois.

Steward was born near Plano, Illinois, in Kendall County on June 23, 1841. His uncle, John Hollister, owned and operated a small workshop. Steward's father, Marcus, and Hollister became early reaper manufacturers. Steward himself became the local "Mr. Fixit" by the age of 14. The Civil War interrupted Steward's machinery career and he served until injuries forced him home. By this time the Steward family had hooked up with the Marsh family to build the Marsh Harvester, the first true advance beyond the reaper and the foundation of the binder. Steward helped manufacture the harvester from 1865 to 1870. The Marshes separated into their own company in 1869.

In 1870, Steward went on the adventure of a lifetime: the second U.S. government expedition through the Grand Canyon. Steward went as an unpaid assistant geologist, which suggests his knowledge went far beyond machines. After five months of exploring the Grand Canyon, he was forced to leave the expedition because of back injuries and pain related to his Civil War injuries. He had to walk out of the canyon because he couldn't even ride a horse due to his pain.

After recovering, Steward returned to Illinois. Elijah Gammon joined the firm, then William Deering. Deering eventually took over the firm, handling the business side and pushing the twine binder. Steward handled the mechanical side and became the superintendent of the Deering Harvester Company. Deering and Steward were close friends; in later years Steward was one of the receivers of secret bonuses that Deering gave loyal long-term employees. Steward became the point man of the attacks against the McCormicks during the Reaper War, both in courts of law and courts of public opinion. He amassed hundreds of pages of evidence and testimony detailing the McCormicks' false claims as to who invented the reaper and many other machines. Steward also put together the huge display of reaper and binder models and information that went to the 1900 World's Fair in Paris, also aimed at refuting McCormick's claims as presented by McCormick Harvesting Machine Company.

Steward often went to great lengths in trying to beat the McCormicks. A proposal was made at one point in the war that the U.S. Treasury should print a gold note with Cyrus McCormick's picture and a reaper on the back to honor his role in inventing the machine and building the economy of the United States. Of course, only U.S. presidents had been on the bills previously. Steward heard word of the plan and went to Washington, D.C., where he presented the secretary of the treasury with another bill to honor a great American who helped to build the economy—Lydia Pinkham, with her little bottle of patent medicine to cure "female troubles." The secretary of state agreed that the United States of America should not advertise for single businesses, and the plan to put Cyrus McCormick's face on money died. Steward also reprinted the memorial book to Robert McCormick, describing his life's accomplishments, including the story of Robert, and not his son Cyrus, inventing the McCormick reaper.

Steward did not shy away from new technology even in his older years. Under his command, Deering Harvester was noted for bringing out new machines years ahead of the competition. He played an active part with the gas engine experimentation mentioned in the text and was designing tractors up until the time of his death in 1913 at the age of 72. Even his tractors, which aimed at light weight and expanding the use for them beyond just plowing, pulling, and belt work, were ahead of their time.

Steward was also an accomplished author, publishing several articles about his Grand Canyon expedition, several articles and books about harvesting machinery (including The Reaper, perhaps the best history of the harvesting machine industry before 1902), and several books about early Illinois history that were well received in their day.

Ed Johnston sits proudly atop his creation next to the lumberyard at McCormick Works. McCormick Works was then a huge operation on the edge of Chicago and had plenty of room for testing new machines. The first McCormick Automower was a rush job built to compete with Deering in Paris. The Automower was probably crated and shipped to Paris immediately after this photo was taken. *State Historical Society of Wisconsin*

What is important about the Automower (besides the engines) is a part of the technology. Prior to this time, energy for driving an implement had either come from flat belts (such as threshers), traction wheels (such as the bull wheel of a binder or the wheels of a mower), or from the human directly (hand rake reapers). Horse-drawn sweep powers did use shafts, but were notorious for the complex mechanisms involved. Ellis & Steward were the first to patent the idea of transmitting power from the gasoline engine to the working part of the implement through gears and shafts—the first power take-off (PTO). Although this PTO could not be disconnected from the transmission (a "dead PTO"), it was the first to avoid the use of the flat belt or traction wheel. Although experiments had been made back into the 1700s with chain and belt PTO, the Deering Automower was the first modern attempt that would lead to today's PTO.

The Deering Automower had faults, mainly the lack of an effective transmission. Instead of having a reverse gear, the Deering had a reversing engine, meaning that the drive had to be disconnected, the engine stopped, the engine started again in the other direction, and the drive re-engaged to reverse. The "dead PTO" would also cause problems. The engine itself would see even further experimentation and would form the basis of the IH horizontal stationary engines, which began production in 1903 or 1904.

McCormick Harvesting Machine Company also had a young engineer interested in gasoline engines. Edward Johnston appears to have been a child prodigy among harvesting machine men. Johnston began experimenting with gas engines in 1896, completing his first engine in 1897. A second air-cooled engine was fitted into a self-propelled carriage in 1898. It had two forward speeds and a reverse.

This Automower was mainly intended to beat the Deering Automower on the world's stage, which it did at the World's Fair in 1900, outperforming the Deering in a field of lucerne near Paris. Nettie Fowler McCormick, widow of Cyrus Sr., was present and looked on with pride. The McCormick Automower's importance extends far beyond this victory. It embodied several principles that became standard decades later. While Deering had a rudimentary PTO system, the McCormick had independent, live PTO, which enabled the Automower to stop traveling while still operating the cutter bar, a feature that allowed it to clear tangles and avoid bogging down. This feature was the cause of the McCormick victory as the Deering Automower bogged down, killed, and could not be restarted. After this failure in front of the judges and many onlookers, the Deering team tried to rescue victory from the jaws of defeat by insulting the McCormick team and attacking the McCormick hospitality tent, trying to eat all of the food and drink the expensive wines. Some of

Edward A. Johnston

For 45 years Edward A. Johnston was the heart and soul of McCormick and then IH technical development. No other man had such an impact on farm equipment technology. Responsible for 171 patents, as well as uncountable other inventions, improvements, and developments, Ed Johnston was a legend within IH.

Johnston started work at the tender age of 14 for the Johnston Harvester Company in Batavia, New York. His grandfather and uncles were probably connected with this company, while his father worked for the McCormick Harvesting Company in Chicago. His first patent at age 19 was for a box-making machine. Other non farm equipment patents included a razor blade sharpener, automatic cigar lighters, and vending machines.

Johnston joined the McCormicks in 1894, becoming an expert (field representative) and then joining the experimental department under R. B. Swift, where he quickly became a mainstay. His first farm equipment patent was for a vertical lift mower in 1897. During this time he began his gas engine and automobile development, leading to the Automower triumph of 1900.

Johnston left the McCormick Company to become the factory superintendent at the Keystone Company at Rock Falls, Illinois, in 1902. There he developed grain and hay harvesting machines, with the patents remaining in his name (rather than being assigned to the company as was the case at McCormick and later IH). Keystone was bought by International in 1905, bringing Johnston back to IH to stay. He began development of tractors and Autowagons at the Rock Falls Works almost immediately, eventually bringing both into production. Autowagons were the first priority, Johnston overseeing the establishment of an Autowagon factory at Akron, where tractor development continued and production of tractors to Ohio Manufacturing's design began. Johnston's own design would reach production after yet another transfer to Chicago. Johnston oversaw the construction of Tractor Works (and probably participated in the actual construction himself), while still developing his design in the famous small building (described by one observer as a barn) and tent across the street from McCormick Works. During this time period, Johnston was putting in 12 hours a day, 7 days a week, and was described as "a powerful man, almost impervious to physical fatigue." His men were impressed by both his physical abilities and technical knowledge, remembering him fondly in later years. He was also known for his ability to magnificently chew out his workers when necessary, without ill feelings resulting in the worker. He pushed his workers to the maximum during these times, expecting excellent results.

After building Tractor Works and becoming its first superintendent (as well as tractor designer), Johnston was promoted to assistant manager of IH's engineering department, and in 1922 he was promoted to manager. As the head of the engineering department, he had responsibility for all of IH's engineering and experimental work, which was then scattered in several dozen locations. He had oversight of tractor, farm equipment, motor truck, and other product development in the far-flung IH empire. Johnston's last promotion came in 1934, that to vice-president of engineering and patents, giving him control of the patent department as well. Before retiring on March 1, 1939, Johnston had overseen the development of the Farmall and PTO, motor truck development, and the cotton picker.

Johnston was an important member of both the Society of Automotive Engineers and the Society of Agricultural Engineers, which presented him with the Cyrus Hall McCormick Award in 1938 for distinguished contributions to farm engineering. Johnston died in 1946 in Santa Monica, California. He had gone to California (as had many other IH retirees) for "health reasons" as he endured great pain from arthritis in his later years. He was probably still doing some consulting for IH at the time of his death.

the McCormick team started to head toward the Deering men with a view of tossing them about (which almost certainly would have resulted in a knockdown fist-fight in front of the crowned heads of Europe), but Cyrus McCormick Jr. stopped his men from proceeding further. The McCormick victory was of great satisfaction, especially to Nettie Fowler McCormick, who wrote back with pride in this accomplishment.

While Deering had a reversing engine, the McCormick had a full reversing transmission. Johnston applied for a patent on these features in 1902, two years after the event. Although McCormick Harvesting Machine had no plans to sell Automowers before the World's Fair, after the victory development of the Automowers continued. A second Automower was built with a single-cylinder engine. Patents were applied for but not granted until 1904, when Harvester clearly had other plans. This patent covered the use of the "live PTO" for agricultural implements.

The second McCormick Automower was demonstrated at the 1903 St. Louis Exposition and then stored. In 1959, the Automower was donated to the State Historical Society of Wisconsin, along with other historical IH machinery and models. The Automower can be seen at the State Agricultural Museum at Cassville, Wisconsin, during the museum's season. The Automower is in poor condition, but it is still believed to be the oldest of the Automowers in the United States and is still worth a visit. The museum also has the first rubber-tired farm tractor, the U that Allis-Chalmers experimented with in 1931.

McCormick Harvesting Machine's manager for Germany, Mr. Vogt, sits on the latest machine from the home office. Note the wires leading from the battery box, which was directly behind the "COR" in McCormick. The front tank held the battery and fuel tank along with the cooling water. Some sources state that the McCormick Automower had two cylinders, but this version definitely has one. *State Historical Society of Wisconsin*

The Consolidation

By the time the IH consolidation occurred, Deering was moving toward the manufacture of gas engines, which it apparently did after the consolidation. Some reports have Deering in production at the rate of 100 engines a day, but no further evidence of this has been found. The reports probably refer to Deering Works production after the consolidation, as it is known that output was begun there before all IH engine production was transferred to Milwaukee. It was Deering-designed horizontal engines that would make up the majority of engine sales and power the Ohio Manufacturing tractor-trucks that would bring International into the tractor business.

McCormick was moving toward producing a gas engine before the consolidation, but the company seems to have been behind Deering. Vertical engine production started at McCormick Works before being transferred to Milwaukee Works, with production starting in 1905. These engines were somewhat similar to the engine in the second Automower.

Although efforts were taken to consolidate upper management, at the works (factory) level things remained much the same. Each works maintained an experimental department devoted to not only the machinery produced by that works, and techniques and production machinery necessary, but also whatever machines its experimenters and factory superintendents were interested in. The general office also had its own experimental department to oversee the work. The separate engineering departments tended to preserve the rivalry between Deering and McCormick after the IH merger.

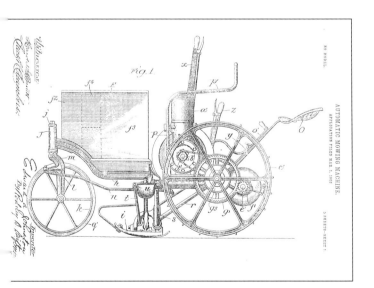

McCormick's version of the Automower shows that the tank up front actually contains the battery (F4), the cooling tank (F3), and the fuel tank (F2). The steering lever is P, while O1 is part of the gear shifting mechanism (a foot pedal not shown is the second part). X controls the clutch for the mower bar drive. Y and Z control the height of the mower bar. *U.S. Patent*

Chapter 2

Early Tractor Development

Ed Johnston left International Harvester at about the time of the consolidation (1902), going to the Keystone Company. In 1905, International Harvester bought Keystone, probably for the machines Johnston had developed. IH was looking at other new product lines and, while still located at the Rock Falls Works (the new name of the Keystone factory) after the purchase, Johnston designed and partially built the first IH tractor, a tricycle type that was never completed. The Johnston experiments (including tractors and auto-buggies) were transferred to McCormick Works in February 1907, and a second tractor was designed and completed, possibly using parts of the first. This tractor was entered in the Winnipeg contest in 1908. A variety of motors were tried, including a Brennan four-cylinder and a three-cylinder of Johnston's design. The three-cylinder was fitted to the tractor after the next transfer of the work to Akron Works in the fall of 1907. A two-speed forward and one reverse transmission was fitted with a friction clutch, with final drive being a malleable chain. Although this tractor was not successful, according to Johnston it was the first step to the Mogul line of tractors. After the return from the Winnipeg tractor contest, design work was started for the Mogul 45.

International Harvester also tried a more proven method. The Morton Tractor-Truck, as manufactured by the Ohio Manufacturing Company, had been in existence for several years and had proven itself with the engines of several manufacturers. A tractor-truck was a frame, gearing, steering gear, and drive wheels of a tractor—everything except the engine and cooling system of a tractor. The purchaser of the tractor-truck then added the engine of his choice, resulting in a tractor of at least minimal capability, certainly better than some of the manufactured tractors of the day. Power was transmitted from the engine to the gears through friction pulleys. IH purchased a tractor-truck and fitted a Milwaukee-built Famous horizontal engine. The first engine was shipped without the IH-built cooling tank, necessitating the manufacture of a cooling tank on site. Fourteen were produced in 1906 "to test the market." The tractor was a success, and Interna-

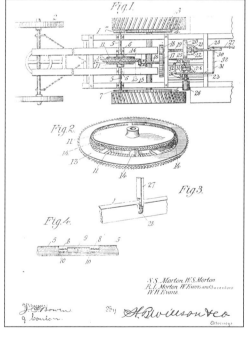

It has been long known that S. S. Morton was responsible for the invention that became the Ohio Manufacturing Company's traction truck, but what was not known is that several members of the Morton family were involved, as was W. H. Evans, in other designs. The design shown in this patent indicates that the Mortons were involved in a gear drive tractor in 1902. This tractor was probably not produced for sale by the Ohio Manufacturing Company. *U.S. Patent*

S. S. Morton

S. S. Morton was one of the real pioneers of the tractor industry. He started in the business in 1899 with the Morton Manufacturing Company of York, Pennsylvania. There he built several hundred tractors. These tractors were chassis to which a stationary engine could be mounted. In 1904 he went to the Ohio Manufacturing Company, where he helped manufacture a friction-drive tractor (again, a truck that a stationary engine could be attached to) that he had developed while at the previous company.

In 1910 Morton sold his interest in the friction-drive tractor to International Harvester. Ohio Manufacturing also sold its interests in the tractor to IH but remained the manufacturer. Morton became interested in four-wheel-drive tractors and formed Morton Tractor Company at Fremont, Ohio. He operated this company with his son, R. B. Morton. The actual manufacturing was done by Heer Engine Company in Portsmouth, Ohio. Morton described this tractor as having a link-chain drive with a gear transmission.

Morton moved to Harrisburg, Pennsylvania, in 1911, and re-incorporated his company as the Morton Truck & Tractor Company of Pennsylvania in 1912. Morton came out with a second four-wheel-drive tractor, this being driven by worm gears, which he stated was sold to allied governments for military purposes in World War I. There was also a four-wheel-drive motor truck based on the tractor design that was sold during World War I.

After World War I, Morton began the development of the gear-less transmission, which he was still actively doing in 1921 in partnership with the Keystone Gearless Tractor Company in Philadelphia, which was later known as the Keystone Tractor and Truck Company. This tractor had no gears or transmission, but it had an infinite number of speeds up to six miles per hour. A crawler version of this machine was also experimented with.

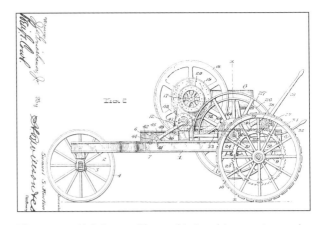

The more widely known Morton friction-drive traction truck is shown in this patent. The whole engine moved back and forth on castors to bring an engine-driven friction wheel into contact with another friction wheel that, in turn, was geared to the drive wheels. *U.S. Patent*

Type A/Type B

The friction-drive tractors quickly proved that a good tractor could be built and sold, but also proved that improvements were needed. Ohio Manufacturing (mostly Kouns, although Morton also participated) designed a tractor-truck with a gear transmission. The Type A and Type B gear drives were originally meant for only light-traction service, pulling threshing rigs, wagons, and other machinery. Plowing was not especially advertised, but the Type As and Bs were quickly put into that service.

The main difference between a Type A tractor and a Type B tractor is the rear axle. Type As only had a half-axle system, while the Type Bs were made with a one-piece axle that carried across the whole width of the tractor. The Type A was produced in 12, 15, and 20 horsepower. The first Type A was shipped in 1907, with

tional entered into a partnership with Ohio Manufacturing. IH referred to Ohio as the Upper Sandusky Works, although IH never actually bought the company.

Although S. S. Morton usually is given much of the credit for the development of the tractor-truck, Moses W. Kouns supervised the refinement of the Milwaukee stationary engines into a tractor engine, including special flywheels and extended crankshafts. The decision to do so was approved by Maurice Kane on December 18, 1906, although drawings and patterns for the special tractor parts had been arriving since November 14. Kouns seems to have been the man at Ohio Manufacturing who IH dealt with the most, and it was Kouns who supervised the next advance.

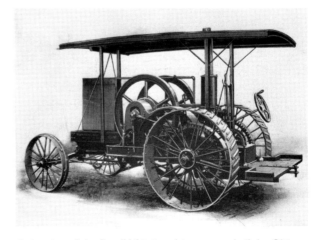

A drawing of the first IH friction-drive tractor built by Ohio Manufacturing Company. This tractor has the Ohio-built cooling tank. Ohio had by this time built several hundred tractors using other companies' engines, but IH would soon take over all production. *State Historical Society of Wisconsin*

In the twenties, IH repurchased the oldest tractor the company could find. This is No. 13, the Arkansas Traveler. The first 14 tractors built were followed closely by IH as part of examining whether to enter the tractor market. They performed well, and IH entered the business. *State Historical Society of Wisconsin*

the first three Type As, two-speed with friction reverse, being shipped in 1909. Type A two-speeds with gear reverse were first shipped in 1910. Type Bs were built in 20 horsepower only, with the first shipped from Upper Sandusky in 1908.

The demand for Type A and Type B tractors apparently outstripped Ohio Manufacturing's ability to produce them, and production began in both Akron and Milwaukee Works in 1908 (although Upper Sandusky re-mained in production until the 1910s). Samples and patterns were sent from Upper Sandusky to Akron. It was quickly discovered that Ohio Manufacturing's methods were a little primitive. Gears were made to fit on each individual tractor and used no standard pitch

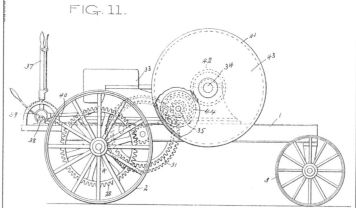

Moses W. Kouns worked for Ohio Manufacturing, and it was he who designed the next step from the friction-drive tractors—the gear drives. However, Ohio Manufacturing's methods were a little primitive and IH ended up redesigning the tractors (at least the ones it produced). *U.S. Patent*

or pitch diameter. Akron Works tried to make the gears to the next closest standard, but when the first of the new gears came through, some gears couldn't be driven into place with a hammer while others had teeth that refused to touch their mates. After several days and a conference involving the management, the Ohio samples were eliminated and International designed a new gear train. A man named Huff was hired to do the drawings under the supervision of Johnston and the direction of H. B. Morrow, Johnston's assistant. J. F. Steward was also heavily involved and may have done the actual engineering of the new gear train. After this experience, all of IH's tractors were made to a set of drawings instead of the older

The first all-IH tractor is shown here at the Winnipeg Tractor Trials of 1908. The tractor had already traveled through three IH experimental departments in two states before even getting this far. Edward Johnston's first effort would eventually become the Mogul 45. *University of Saskatchewan, H. A. Lewis Collection,*

sample system. Akron shipped a tractor-truck to Milwaukee, which made up a new list of drawings and specifications. According to L. B. Sperry, who eventually became International's chief of tractor engineering, Akron produced tractors to be shipped east, while Milwaukee produced tractors that were to be shipped west. Apparently, Upper Sandusky shipped tractors everywhere, although those intended for export were shipped first to Milwaukee. Milwaukee changed even more of the design, especially the 20-horsepower versions, which received heavy reinforcement of the frame, a different friction wheel for the reverse, moving the chain roller brackets, enlarged and strengthened bull gear, an increased capacity cooling tank, and a different brake system. Those were just a few of about a dozen changes.

Type C/Mogul Type C

The Type C tractor-truck was developed in response to the Type A and Type B being put to plowing duty. This tractor was designed by the Akron Works Experimental Department led by Ed Johnston and designed from the start for plowing. Reverse was still friction but the forward direction was now all gear-powered. Larger drive wheels were used, as well as heavier construction throughout, especially the frame. The first few were shipped in 1908 from Akron before production was transferred to Milwaukee Works. Although built at Milwaukee, the Type C later became

The 40-horsepower IHC is on the far right of this photo, clearly demonstrating the low-slung nature of the design. The 40-horsepower performed well, having plenty of power, but water consumption was very high. The other competitors (from left to right) are a Marshall 12 horsepower, an IHC 15 horsepower, a Kinnard-Haynes Flour City 40 horsepower, an IHC 20 horsepower, a Gas Traction 25 horsepower, and the IHC 40 horsepower. *University of Saskatchewan, H. A. Lewis Collection*

known as a Mogul. This tractor was built in kerosene and gasoline versions and was the winner in its class at the Winnipeg tractor test in 1911.

Designed at Akron and built in Milwaukee, this was the first all-IHC tractor in production. This tractor was built in the 20-horsepower size until a 25-horsepower joined the line in 1911 (production was authorized

The Mogul 45 on the streets of Chicago, June 11, 1910. The early Mogul 45s (the first 50) were built in a small building and tent next to IH's huge McCormick Works, while the new Tractor Works was being erected. As shown here, testing was done in the streets. The 1910 Moguls had the fan on top of the radiator and were in general of lighter construction than the tractors that would be built in Tractor Works starting in January 1911. *State Historical Society of Wisconsin*

December 20, 1910, using the standard Milwaukee Works 25-horsepower stationary engine with a base designed for mounting for traction use). It is interesting to note that the Type Cs were the only Moguls built in Milwaukee Works with the small exception of a few Mogul 8-16 engines built during the height of popularity for that tractor. The Type Cs were changed into kerosene tractors August 24, 1911, although the governing was not changed from hit and miss until September 15, 1913, when it was finally changed to throttle governing because it was thought that this was the only way to sell the remaining tractors.

Chicago Works Large Tractors
Mogul 45 and Mogul 30-60

In Akron during 1908, the Experimental Department under Johnston started actively designing and testing a new 45-horsepower tractor, which was to become the Mogul 45. This was one of the first IH production tractors with an engine specifically designed for the tractor, although the engine design was used in modified form for stationary engines (the Mogul "Giant" engine, rated as 50 horsepower). The first design had two gear speeds forward and one friction reverse speed. The original motor had a 9x12 engine that was later increased to a 9 1/2-inch bore. According to one account, the first Mogul 45 was completed in fall 1909 and was tested for a short time in Akron. It was then sent to the swamps near Kankakee, Illinois. The Kankakee swamps saw many experimental and developmental IH tractors in years to come.

At this time, the Akron Works experimental department was also testing the Type C tractor-truck, which would later be manufactured at Milwaukee. Akron Works, manufacturing the new auto-buggies, had mainly smaller machine tools that were inadequate for tractor work (an employee transferred into the tractor experimental department said that one casting for an experimental tractor looked as large as an entire auto-buggy). Most of the machining for the experimental work was done by an outside contractor, Welver-Seaver-Morgan Company, until the move to

Chicago. Another incident concerning Akron Works was a spy sent in from the Russel Company. The spy, J.W. Bradley, was assigned to testing tractors. For this rather low and dirty job, he wore a standup collar and a goatee, and was remarked to be a "nice appearing and well-educated fellow." After about three weeks, the fellow he worked with (R. W. Henderson, himself a former Russel employee) discovered that Bradley worked for Russel, with the obvious result of Mr. Bradley being terminated from his position. The history that this information is taken from exists in two versions. In the uncensored version, Henderson stated that Bradley taught him more in those three weeks about tractors than he had ever learned from any other IH employee!

Ed Johnston was brought back to Chicago in 1910 and made superintendent of the new Tractor Works, which was to be situated next to the McCormick Works in Chicago. The only problem with this idea was that as of yet, there was no Tractor Works. Construction of a modern factory was begun, but at first Tractor Works consisted of a small building, a tent, and a 35-horsepower engine driving a few machine tools. Most of these machine tools were taken from McCormick Works' junk pile and reconditioned by Tractor Works' machinists. While working in a tent may have been an advantage in the summer, it was a memorably horrible experience in the winter. Some processes, such as the assembly of wheels, were actually done in the great outdoors without even the protection of the tent. Other problems were caused by the size of the parts. Mobile cranes were used to move the parts, but the cranes occasionally tipped over. Luckily for all involved, the factory was in use by January 1911. At first some of the building stood empty, but this would soon change.

A major component of Tractor Works was, of course, the Experimental Department. After the transfer to Chicago, Johnston completed the development of the Mogul 45. Production started in the temporary factory on Blue Island Avenue in front of the McCormick Works. The Mogul 45 quickly became known as a good, powerful tractor. Fifty tractors were built in the small temporary factory (they built an addition to the factory by buying a circus tent) before Chicago Tractor Works opened in January 1911. The tractor was not perfect: several areas needed strengthening due to the high stresses of large implements and a pounding hori-

The Deering South American Tractor (also known as the Deering 30-60) was probably the most powerful tractor IH built for decades. Accounts of the testing state that this tractor was more powerful than any other IH tractor of that time, presumably including the Titan and Mogul 30-60s, which were tested alongside this tractor. A huge two- or three-cylinder upright engine located in the cab itself provided power. A revolving flywheel without a guard in front of the driver, along with extreme vibration, made this tractor difficult (and nerve-wracking) to operate.

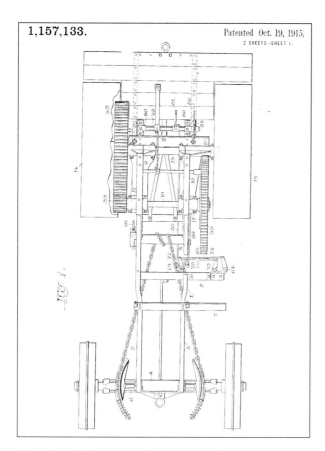

Milwaukee Works patented its own gear drive system, applying for the patent April 30, 1910. By the time the patent was finally granted, the system was nearly obsolete and would be produced only two more years. Henry A. Waterman, the superintendent of Milwaukee Works, was listed first on this patent. *U.S. Patent*

zontally opposed engine. Ten additional experimental Mogul 45s were built in 1911, with revised brakes, oiling system, rear axles and differential, and other more minor changes.

The tractor was changed extensively again in 1912. In August another 10 experimental Mogul 45s were built for test that fall and possible 1913 production. The changes included using two Mogul Jr. cylinders (which had a 10-inch bore instead of the 45s 9 1/2). The Mogul Jr. cylinders had been developed from the Mogul 45 originally, so the cylinders had come full circle. Changes also included an enclosed cooling system, larger bull gears, larger diameter front wheels, a better water pump, and a yet further strengthened countershaft and rear axle to withstand the increased horsepower. These changes proved themselves and were placed into production. In November 1912, the Mogul line received new horsepower ratings for each tractor, with the 45 being rerated to 30-60. This tractor was tested extensively during spring 1913 in Kansas, along with the Mogul 12-25, the Four-Wheel-Drive, the Waite and Steward tractors, the South American tractor, and regular production IH tractors.

20 Horsepower on Type C Trucks

Other than a brief mention in a parts manual and what may be a photograph, no records exist for the 20 horsepower on Type C trucks. There were only nine built according to the parts manual, all in 1911. This tractor may have been a way to test a new Mogul engine for a smaller tractor, the Mogul Junior, using ready-made trucks from Milwaukee. Tractor Works Decision 35 deals with a 20-horsepower kerosene or gasoline engine for traction or stationary use. It states that many of the parts in the engine are regular Milwaukee Works 20-horsepower engine parts and that the Milwaukee Works accessories are used on the engines. Mounting dimensions were also identical. Reportedly, one of these tractors still exists in Australia.

Mogul Junior 25 Horsepower

The Mogul Junior 25 Horsepower was developed by taking the large engine design of the Mogul 45 and installing only one of the two cylinders (10-inch bore instead of the 45's 9 1/2-inch bore), along with other changes designed by Johnston and C. I. Longnecker, who would become a mainstay of IH tractor engineering. The decision authorizing construction also notes the engine could be used as a stationary engine by adding a subbase. The first Mogul Junior was shipped in 1911, although in much smaller numbers than the 45. The Mogul Jr. was extensively changed in 1912, like its larger brother.

Mogul 15-30

The Mogul 15-30 was the old Mogul 25 Horsepower Junior rerated during the horsepower rerating of IH tractors in November 1912. The tractor was also changed to throttle governing to burn kerosene and saw changes for the 1913 model year as well.

Mogul 10-20

Not to be confused with the later more popular lightweight tractor, the first 10-20 was a heavyweight based on the Mogul Jr. design but scaled down to 20 horsepower (it was originally known as the 20-horsepower Mogul Junior), while the original Junior was rerated up to 30 horsepower. First shipped in 1913, this tractor was produced only until 1915, a victim of the trend toward light tractors. A comparison of the old Mogul 10-20- and the new Mogul 10-20 dramatically demonstrates how much smaller and lighter (and therefore cheaper) the later tractors were. The Mogul 12-25 was introduced only four days after the 10-20, probably removing many potential sales.

The Mysterious Mogul 20-40

International Harvester's production records indicate that two Mogul 20-40s were produced in 1914, with some design work definitely occurring in 1913, although work on a four-cylinder vertical tractor was done in 1910 and apparently dropped. In 1915, eight Mogul 20-40s were produced before the tractor was discontinued. An IH photo album contains a gap in which it is written that the pictures for the 20-40 were given to the IH Patent Department in 1931 because the pictures show the 20-40s "one-piece" frame. Ed Johnston stated in a different history (*Evolution of a Tractor*) that the 20-40 had a four-cylinder horizontal engine mounted directly on the transmission case, which was also the frame of the tractor. Johnston also stated that as near as he could determine, the Mogul 20-40 was the first of the integral frame tractors (he only refers to the "general construction of the Ford, Samson, Twin City, and others"). The decline in demand for large tractors was the reason for the demise of the 20-40 after only 10 copies produced.

The first enclosed, one-piece framed tractor is usually believed to be the Wallace tractor produced in 1913 by the J. I. Case Plow Works (later purchased by Massey-Harris). This is seen as a major technological advancement, as the unit frame (also referred to as a boiler plate frame) eliminated the use of steel girder frames and increased the rigidity of the frame. This new frame provided a structure that gave more protection to moving parts (engine and gearing) from dust and dirt. The unit frame was a major step on the way to today's modern tractor. It is interesting to speculate that IH may have been the first with this technology so often attributed to Wallace. It was 1921 before International again produced a one-piece frame tractor in the shape of the International/McCormick-Deering 15-30.

Deering Tractors

Another limited-production tractor is the Deering South American tractor. This tractor does not show up in the IH tractor production figures, and only four photographs are known by the author to exist of this tractor. Recollections of some of the early engineers indicate that this tractor was created to meet the demand for large tractors that existed in Argentina, which was an important market for International Harvester. The existing photographs and reminiscences indicate a tractor very different from the rest of the Titan and Mogul lines. The engine is vertical and apparently based on the two-cylinder vertical stationary engine from Milwaukee, although one of the reminiscences has the tractor engine being a three-cylinder. The early IH vertical two-cylinder stationary engine indeed was designed so that it could be produced in a three- or even four-cylinder version, although apparently this was never actually done for stationary engine production.

Production of the Deering South American tractor was somewhere between 7 and 11 tractors, with all but the first being sent to South America (the first may have gone there after testing as well). Offical rated horse-

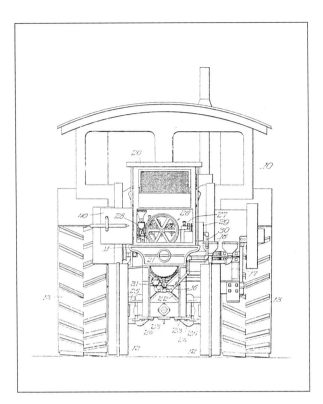

Milwaukee Works also designed a starting system that was quite different from what the Chicago Tractor Works was building. The system used a small air-compressing engine located in the radiator compartment to build up air pressure (stored in tanks under the main frame), which in turn would be used to turn the engine over. A very advanced system for the day, but also more expensive and heavier than the Mogul friction-starting pony engine, and the air system was dropped in favor of the Mogul system. *U.S. Patent*

power was originally 50, being rerated to a 30-60 at the same time as the Mogul 45 was rerated to 30-60 (November 1912). However, IH did underrate tractors to make them fit into a rating class (as happened with the Titan 15-30, which was derated to 12-25 to match the Mogul 12-25), and one traveling expert called this the most powerful IH tractor produced. It is possible that this tractor had many more horsepower than the rating suggests.

That expert, R.W. Henderson, had experience with a tractor sent to Argentina. Henderson stated that when the tractor was run, it was practically impossible for the operator to stand on the platform for more than two or three hours. Given the times and the total lack of sympathy for operator fatigue (engineers in threshing crews were putting in 18-hour days at that point), the vibration must have been extremely bad. Other than the vibration, the tractor was trouble-free, according to Henderson. While riding in the cab of the Deering tractor, Henderson said he shot a bushel basket of partridge on the first two trips across the field. The birds must have been thick because his aim couldn't have been good with the vibration.

Photos of early prototype tractors are hard to identify at this date, mainly because few of the photographs have captions, and also because the photos from the Milwaukee and Tractor Works experimental departments have largely disappeared. However, this is a photo from the Milwaukee Experimental Department, and a different photo of this tractor shows that the engineers labeled the front bolster of the tractor in chalk! This is the one and only Titan 30 horsepower being tested near Milwaukee. This tractor bears a strong resemblance to the production version, the Titan 18-35. *State Historical Society of Wisconsin*

Milwaukee Works Large Tractors

Milwaukee Works was involved with tractors from the first IH tractor until closing in the early 1970s. After the International Harvester consolidation, Milwaukee Works was IH's precision "high-tech" factory and remained so for many years until the operation became largely a foundry. Production of the old Milwaukee Harvester Company's line of binders and other harvesting machinery was transferred to McCormick Works in Chicago in 1905. Existing engine production at Deering and McCormick Works was transferred into Milwaukee, consolidating IH's engine production in one plant (at least until Auto Buggy production started). Milwaukee started production of the Deering-designed engines in late 1904, and experimentation there quickly expanded the line in types, fuels, and horsepower. McCormick-designed vertical engine production started in 1905.

When IH decided to try the Morton tractor-truck as produced by Ohio Manufacturing, it was Milwaukee that supplied the engine and provided all training, technical support, and shipping services for the salesmen and branches involved in the tractor and engine business up until the involvement of Akron. Milwaukee Works remained a factory that was deeply involved in technological advances until the combination of the IH tractor and engine engineering departments in 1917. Milwaukee was the site of IH's first diesel engine mass production in 1933. IH's stationary engines would be built exclusively in Milwaukee until 1911, when Chicago Tractor Works started the Mogul engine line to keep the factory busy when not producing tractors.

As mentioned, Milwaukee maintained a completely separate engineering organization until 1917. This engineering organization had as much, if not more, talent than the competition (meaning Johnston's team) and was almost certainly started by Ellis and Steward as an extension of the Deering efforts in engine production (Johnston was working for Keystone at the time). Tractors were sold under the International Harvester name, then briefly "Reliance" in 1909 for Milwaukee tractors. In 1910, the names were changed on engines and tractors, with Titans sold by Deering dealers, while Moguls were sold by McCormick dealers.

Type D Tractors

At first Milwaukee Works tractor work was limited to the production of motors for the Upper Sandusky and Akron produced tractors. L. B. Sperry later recalled the early Milwaukee tractor activities. During the early spring of 1908, the Milwaukee Works experimental department began experimentation with tractor-trucks powered with Lambert and Brown engines. Around the first of June, P. R. Hawthorne was hired to design a new tractor from the bottom up. The summer of 1908 was spent in design and layout and further tests on the Brown engine, which was finally rejected for weight and lack of power. By fall, Hawthorne began designing a twin-cylinder twin-crank engine, which had a throttling governor, kerosene-burning equipment, removable cylinder sleeves, a pressure-tight crankcase, and a transfer port to the cylinders that pressurized the fuel/air mixture. The chassis was all gear-driven with gear reverse. Automobile steering was used. Many of these features, of course, were years if not decades ahead of their time. Testing of the Milwaukee Works 45-horsepower tractor began in early 1909.

In May 1909, according to Sperry, a decision was made to develop tractors similar to the Hawthorne tractor in 20 and 25 horsepower for the Winnipeg plowing contest in early July. May 22 saw the actual definition of the program and start of design work. Sperry was brought into his first tractor work at this time, being transferred from Milwaukee's stationary engine department to design one of these tractors. Forty days later, during which Sperry was night foreman of the tractor assemblers, the tractors were completed, given a short trial, and shipped to Winnipeg. Sperry said these two tractors didn't win any gold prizes, which is hardly surprising. In fact, the large Hawthorne tractor (then rated at 35 horsepower) broke down during the contest and had to be withdrawn.

After Winnipeg, the 35-horsepower Hawthorne tractor was redesigned into the lower tech but more accepted hit-and-miss governed engine without the transfer port that pressurized the intake found on early Type D tractors, with completion of the redesign around November 1, 1909. The smaller versions of the Hawthorne tractor were redesigned and samples of all three versions were made and shipped to Texas and Oklahoma for testing in the winter of 1909-1910. Known as the Type D series, these tractors were first authorized for production in all three sizes on May 25, 1910. Originally known as Reliance tractors, these tractors were renamed Titan when the Mogul and Titan lines were separated. The naming of the Mogul line (then Type C tractors) took place January 25, 1911, while the Titan name was given to the Reliance Type D tractors in December 1910.

The D line was converted into kerosene tractors due to demand in 1911. The decision authorizing the conversion (dated July 13, 1911) notes that the tractors would be regularly built as kerosene tractors and that gasoline tractors in the field could not be converted without the aid of an expert familiar with kerosene tractors. New mixers, air heaters, and fuel tanks were necessary for the new tractors.

The Titan 45 got an entirely new kerosene mixer that dramatically improved horsepower. The decision authorizing the new mixer, dated January 12, 1914, stated that because of the new mixer, the tractor was rerated, becoming the Titan 30-60. All throttle-governed 45-horsepower engines could be fitted with the new mixer, which was probably fairly popular with the dealers because of the increased horsepower.

Road Rollers

The Titan Type D line would also see the first purpose-built power road equipment produced by IH. Although IH marketed all of its early tractors to road building and municipal buyers to meet America's crying need for better roads (if you could call those muddy paths "roads"), the Titan Type D 20- and 25-horsepower tractors were equipped with a road roller front axle. The roller could be interchanged with front wheels and was designated the "convertible." Although these tractors were marketed in very small numbers, they demonstrate IH's commitment to the construction market well before the Industrials and TracTracTors came into the product line in the mid-1920s. It should also be noted that International distributed equipment to markets other than farmers, including railroads, governments (especially townships and large cities), and others. The road rollers were authorized for production May 22, 1912. The construction of the front roller was changed for 1913 from a cast steel yoke running over the roller to a horizontal framework around the roller, which made conversion and steering much easier. The earlier road rollers were built with combination gasoline and kerosene engines with hit-and-miss ignition, which was officially changed to the more effective throttle-governed kerosene engine November 5, 1913 (but actually had been built since September 1).

Titan Type D 30 Horsepower and 18-35 Horsepower

This tractor was basically a Titan 45 with a different, smaller engine, transmission, and automobile steering. The Titan 30 horsepower was designed by E. D. Eliassen under the direction of H. A. Waterman, the Milwaukee Works superintendent and engineer. Design work began in 1911, with design and building of a sample extending into 1912. The Titan 30 horsepower was authorized for production December 27, 1912. The tractor rating, as well as the designation, was changed to the Titan 18-35 on July 19, 1913. The first production 18-35s were shipped from Milwaukee Works in 1913. The design and appearance were similar to the Titan Type D 45 tractor and it was designed to be used with an air starter. While regularly built as a throttle-governed engine, it could be specially ordered as a hit-and-miss.

Steward Experimental Self-Propelled Grain Harvester

John F. Steward did not fade from the scene while Milwaukee and Chicago Tractor was getting under way. Steward experimented with a self-propelled grain harvester in 1909. This machine was a

One of the Steward self-propelled binder experiments. J. F. Steward was a pioneer in mating the internal combustion engine to harvesting machinery, starting with the early Automower and corn harvester experiments, and continuing with this combination of an IH gas tractor and a Plano binder. Two power take-off shafts can be seen running from gearing under the toolbox (under the flywheel) to the binder. Further development of the power take-off (PTO) is covered in chapter 4. *State Historical Society of Wisconsin*

header-binder (a grain header with a binder attachment ordinarily propelled by several horses pushing the binder from behind, used mainly on large grain farms in the West) butted up rear to rear with a tractor, probably a Type A or Type B. which ran backwards. Steward was even then trying to find how to use tractor power to do farm work other than plowing or belt work. The push binder was a good place to start because of several factors: large horsepower was required, the position of the horses put the horses into the dust created by the binder, and the use of the binder during the hottest days of the summer put the horses into even greater distress.

Steward's tractor binder originally had a cut of 16 feet, that being the limit of what the binder attachment could handle. The machine was tested during two harvests in 1909, with the machine cutting 50 acres per ten-hour day. The next stage of development saw an ordinary widecut binder hooked up to the tractor, giving a total cut of 23 feet and 100 acres per day. The limitation to the size of the cut were the binding attachments. If the headers were used without the binder attachments (the heads being dumped into a wagon directly), Steward stated that there was no limit to the size of the header.

The Kerosene Tractors

IH was caught unprepared by the sudden popularity of kerosene tractors. IH's tractors had been designed to burn gasoline and other light fractions of oil. Harvester was forced to sell Rumely Oil Pulls in those markets that did not have Rumely dealers but demanded kerosene tractors, especially overseas. This situation could not be allowed to last for long, and IH's engineers in both the Mogul and Titan lines moved quickly to modify their existing designs in both tractors and engines. What resulted had the same names, had most of the same parts, but had critical differences that in some cases led the engineers to release the new kerosene tractors as being new designs for production rather than simple modifications.

The major change in the new tractors was the governing. Except for the prototypes of the Titan D series, all IH tractors had hit–or miss governing. With a hit or miss engine, if the engine speed is too high, the engine's exhaust valve was held open. Since the intake valve was opened by the suction of the engine, if the exhaust valve was open the air was sucked through the exaust valve, the intake valve stayed closed, no fuel entered the cylinder, and the engine did not fire on that revolution. The engine did not fire again until the engine RPMS had decreased to a certain level. When the slower speed was reached fuel was admitted and the engine fired, resulting in a higher RPM and withholding the fuel again. The cylinder would miss several firing cycles in between firings. However, not having the engine fire resulted in the cylinder cooling down. In a kerosene engine, there low tempetures caused the kerosene that had gone through a mixer to recondense. The condensed kerosene did not burn properly, causing loss of power. The condensed kerosene also tended to leak past the cylinder rings and into the engine oil, diluting it and causing lubrication problems that resulted in more frequent breakdowns and overhauls.

To correct the problem, IH turned to the more modern method of throttle governing, where, if the engine speed was high, a smaller amount of fuel/air mixture was admitted to the chamber, resulting in less power and lowering of RPM. If RPMs slowed too far, more mixture was admitted providing more power and RPMs. Firing every cycle resulted in higher, more even cylinder temperatures and better burning of the kerosene.

Soon every IH tractor would be ready to burn kerosene, with some tractors able to burn distillate, solar oil, and several other fuels. This great switch started in mid 1911 with the Titan agricultural tractors, and then swept through the Moguls. The Ohio Manufacturing Company requested kerosene burning engines for their tractors as well. IH was even forced to convert tractors already in inventory to kerosene in order to liquidate them; gasoline tractors were just not saleable by the mid–1910s.

Chapter 3

Light Moguls and Titans

By 1913, the shift was clearly heading toward a lighter tractor for a variety of reasons. The first was market saturation: Those who could afford and use the large tractors either had them or were waiting for something better to come onto the market. The need for large tractors themselves was actually declining, as the difficult task of breaking the prairies was rapidly nearing completion. Using the large tractors on broken land was difficult; the largest tractors were so heavy that they sunk even in dry land and compacted the soil when they retained flotation. Arranging implements to use the large horsepower meant awkward squadron hitches and large strings of implements. Most farms were not large enough to merit a large tractor. Most farms were arranged for horse farming, with smaller acreage that could be maintained with the labor needed at slow-speed farming. Some farms were so small that a large tractor and string of implements probably couldn't be turned around. A vast number of these farms existed. Another issue in tractor design was cost. Steel and iron are sold by the pound, and every pound of material in a tractor costs a certain amount. Basically, IH could lower the weight and produce a tractor that is cheaper to manufacture.

The shift toward the light tractor was not as simple as designing a small version of the large tractors, as IH had done in producing single-cylinder versions of the two-cylinder tractors. Light tractors were in fact much more difficult to design than their predecessors. Instead of a huge cylinder with low-compression ratios, smaller cylinders with high-compression ratios were necessary, along with higher piston speeds. In the high dust environment of farming, wear became a major issue and shielding and filtering were now needed, further taxing the engine. Transmissions were needed that could handle the higher engine rpm, as well as provide a range of forward and reverse speeds instead of one or two forward and a friction reverse.

The entire tractor had to have better engineering to withstand the wear and tear of farm use. With the heavyweights if a weak spot was discovered, adding metal in that spot to produce a stronger part was the common cure. Starting with the first lightweights, design would increasingly play a larger part in eliminating problems.

In 1913, small tractors first hit the market in a big way. The Bull tractor is often credited as the tractor that started the movement towards smaller machines. Although the Bull was the first small tractor to be commercially successful, Harvester, like most of the tractor

The Steward Tractor during a moment of rest. Steward was a constant promoter of lighter tractors and put his inventive talents where his mouth was in IH's first four-cylinder tractor. The controls for the rear seat are clearly visible in this photo, but the control could be transferred to the engine end of the tractor as well. Steward is barely visible to the left. *State Historical Society of Wisconsin*

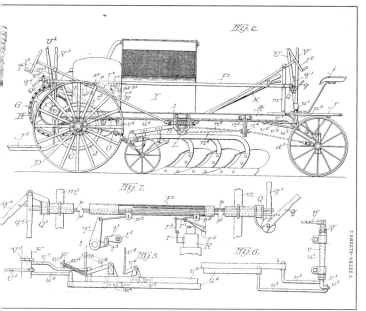

The Steward Tractor as patented. Figure 2 is a side view as it would appear in the field. Figure 7 shows the complex control system necessary for reversible control of the engine and tractor, Figure 5 shows the clutch and transmission control system, while Figure 6 shows another view of the clutch lever and stand. *U.S. Patent*

manufacturers, had efforts in place to reduce the size of its tractors in 1912, and there is a brief mention of a small four-cylinder tractor being experimented with in 1911, which was possibly the first Steward tractor. John F. Steward was a champion of the small tractor. Steward sent a letter to Alexander Legge on August 5, 1911, saying, "Every builder of tractors is in a rut. Tractors have the same work as steam traction engines, and inventors have naturally inclined to follow along in these heavy lines. What we need to do is get out of these heavy lines!"

Deering Experimentals

It should come as no surprise that IH tractor experimental work was active in other works besides Milwaukee Works and Chicago Tractor Works. J. F. Steward was approaching the end of a long, active career by 1910. Among a host of other duties, he had participated in Deering's early gas engine and Automower development. Apparently he never lost his fascination with self-powered machines because in 1911 or 1912 he designed and had Deering Works' experimental department produce, a series of experimental tractors identified under his name. The Steward tractors were an attempt to better mate the tractor's power with the implement. A four-cylinder engine, the first in an IH

Another candidate for the Waite Tractor. This photograph was among glass negatives from the time of the Waite tractor experiments. Although similar in appearance to the Steward tractor, this tractor is much heavier and nonreversible. The engine position is also completely different from the Steward. Although this tractor cannot be identified as the Waite with 100 percent certainty, this is the best candidate. *State Historical Society of Wisconsin*

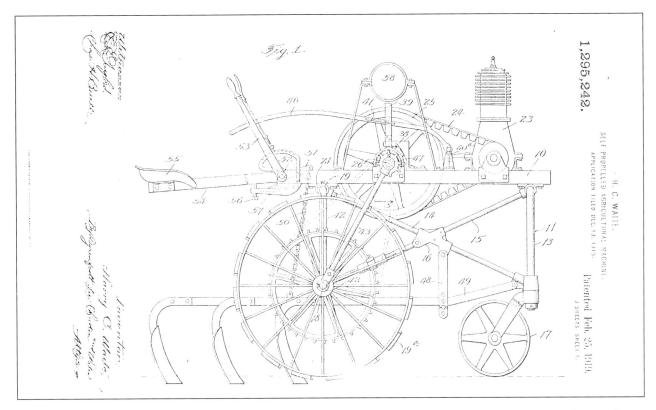

Waite remained in the tractor business for years after his experiments with IH. Here is another of his designs, this one for a small, lightweight motor cultivator. The interesting part of this invention is who he assigned an interest (share) of his patent to: B. A. Kennedy, a top IH official of that time. Was Kennedy thinking about going independent? Or was IH taking an interest in a different motor cultivator in 1915, but through Kennedy instead of more formal channels? *U.S. Patent*

tractor, was cross-mounted at one end of the tractor. The operator could operate the tractor from either end. Implements were mounted beneath the tractor, much like the later Farmall. Plows were especially experimented with.

The other tractor that Deering Works experimented with was a tractor named after its designer, Harry C. Waites. Waites outlined his ideas in a letter to B. A. Kennedy in 1913. His tractor was to be lightweight, weighing less that 5,000 pounds, selling for $1,200, and sufficient to do the work on a farm of 200 acres. Two Waites tractors were built and one was sold. It is not specifically known what the Waites tractor looked like as no photographs exist that are definitely labeled "Waites Tractor." Waites did patent a tractor with an assignment of rights to the same B. A. Kennedy as mentioned above. Kennedy was a top executive at this time, which leads to the conclusion that either that particular tractor is the IH tractor, or Kennedy was moonlighting. Waites was one of those entrepreneurs whose work abounded during the early years of the tractor industry. He worked for several companies, sometimes at once, designing tractors of various designs. Waites is known to have worked for IH and Elgin Tractors, and probably several others. Waites

This is probably the first Mogul 12-25 being tested near Chicago Tractor Works. IH revised the muffler system several times on these tractors. The sheet metal on this version (the hump next to the cab didn't make it to production) is quite attractive. The production version of the sheet metal was probably cheaper to produce. *State Historical Society of Wisconsin*

The only photograph of the four-wheel-drive Mogul 12-25. IH designed several innovative tractors in 1913, but none would make it into production. IH would develop production versions of these experimental ideas years and even decades later. The tractor industry was changing rapidly during 1913 and 1914, and all ideas were on the table. *State Historical Society of Wisconsin*

An undated photograph, this shows an early Mogul 8-16 with the sheet metal water hopper (rivets are seen on the hopper). This photograph may be of the prototype, but the lack of a date means that it is only a possibility. *State Historical Society of Wisconsin*

later formed his own company to sell a tractor under his own name, with a Waukesha engine and assembly in Waukesha.

Milwaukee Works Light Tractors
Titan 15-30

The Titan 15-30 was to be the Titan equivalent of the Mogul 12-25, although the Titan 15-30 was produced in greater numbers and years. The tractor was built with spring mounting, automobile-type steering, a fan-cooled radiator, and other modern advances. The 15-30 rating caused a problem, as the tractor had higher horsepower than the older Mogul 12-25. IH's answer to the problem was to rename the tractor the Titan 12-25 so the Mogul could continue to be sold at the same price.

This tractor was powered by IH's first four-cylinder production tractor engine, although the engine was still horizontal and not vertical. Design started in the summer of 1912 with an L-head engine. Compactness and protection against dirt were emphasized. This tractor was completely spring mounted. The first sample was tested on April 14, 1913, but for some reason Milwaukee Works hid this fact from the rest of the company until mid-August. Production of the 15-30 was authorized June 16, 1914, but the production records indicate that only 13 were produced in 1914 (these may have been for testing). The decision authorizing production makes the unusual note that "tractors may be had in limited numbers about 20 weeks after approval of this decision." While the giving of a date

available is not unusual, the notation of "limited numbers" and the long lead time is very unusual. Even more strangely, a handwritten note corrects the decision to read "after receipt of manufacturing order after approval of this decision" indicating that the 15-30 project may have had opposition within IH's upper management or sales department.

The first Titan 15-30 prototype was sold in August 1914. The tractor had received three poor paint jobs by then and the new owner was convinced that he was sold at least a second-hand tractor. After the enraged farmer had been calmed, and the tractor had proven itself, he convinced three of his neighbors to buy the production versions.

This tractor would receive a redesign with a new transmission and valve-in-head motor in 1916 and be rerated back to a 15-30. Milwaukee Works eventually designed the enormously popular Titan 10-20, production of which severely pressed Milwaukee Works' capacity. The 15-30 was transferred to Chicago Tractor Works to make more room. Tractor Works redesigned the 15-30 to a limited extent, which included new steering knuckles, a new fan shaft, and the change of the serial number prefix to EC (Chicago production started with EC-501, although an experimental tractor may have been built before this). The decision to transfer production was made August 4, 1917. With the change, the tractor name was change from the Titan 15-30 to the International 15-30. After 500 International 15-30s were built in Chicago, the cab was eliminated.

Titan 10-20

The Titan 10-20 was the Milwaukee Works light tractor considered as the tractor that firmly put IH into the tractor business. This tractor equaled the production of all other IH tractors combined until the end of production in 1923.

Design of the Titan 10-20 began in the late fall of 1914. The first sample was field tested in April 1915. Two more prototypes were built and tested through the summer and fall of 1915 near Milwaukee and then sent to Texas for winter testing. A move was made to authorize production of this version in September 1915, but the tractor was not approved. Former chief engineer L. B. Sperry said that 15 tractors of the same early design were built and tested the next summer of 1916 (this was probably 1915). Sperry stated further that the Titan 10-20 was redesigned in November 1915 by separating the cylinders from the crankcase, adding a high-tension magneto and improving the transmission. This was the form in which the tractor was placed into production, which was actually authorized January 22, 1916. IH records indicate that seven tractors were produced in 1915. The various dates and confusion were probably caused by the November 1915 redesign.

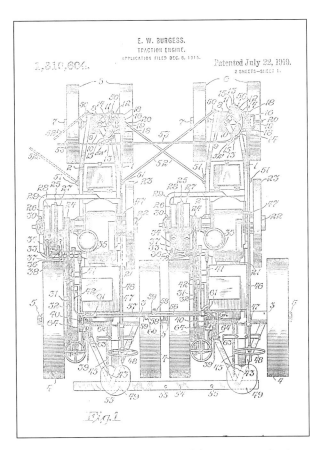

Although the small tractor dominated the tractor market in 1915, there were still some farmers who needed a large tractor. Here is one attempt to meet that market with two smaller tractors, which would have been more economical than maintaining manufacture of just a few of the old large tractors. The two tractors involved in these patents are Moguls. There is no record of production of this system, although several manufacturers tried the idea over the years. *U.S. Patent*

One of these early Titans had caused quite a stir during a stunt at Carlinville, Illinois, when it plowed nonstop for 60 hours.

Milwaukee Works made a promise to place the tractor into production by March 1916. Apparently, difficulties were encountered as the first production Titan 10-20 was produced on February 28, 1916, with many parts borrowed from the Milwaukee Works experimental department. The Titan 10-20 quickly became the most popular IH tractor produced and a name well remembered today.

Major changes occurred to the Titan 10-20 on November 1, 1919, when a higher-speed 10-20 was authorized. Engine rpm were increased to 575 from 500, increasing road speed as well. New lightweight pistons were used along with counterweights. The tractor decision letter stated that the tractor serial number letter prefix had to be changed (from TV to TY). Another change, the enlarging of the water tank to help the

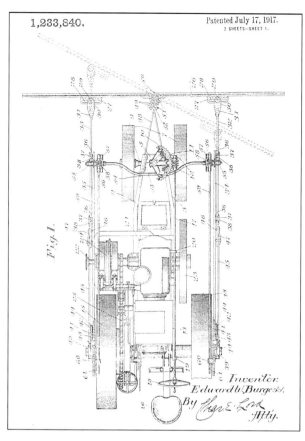

IH remained interested in the road-building business even after the Titan Road Rollers failed to sell well. Here is another attempt by E. W. Burgess to do something unusual with Mogul small tractors, this time a road grader. *U.S. Patent*

engine cool at the higher rpm and power, was proposed but never approved.

The Mysterious Titan 6-12

One IH production record (a penciled-in draft) indicates 19 Titan 6-12s were sold in 1916. The author was unable to find any other information or photographs of this tractor. The best guess is that IH considered a smaller class of tractor under the 10-20 size. This would explain the Mogul 5-10, which was not mentioned in the report as it was for sales only. The production run of 19 would indicate a preproduction tractor sold to selected farmers for on-the-farm testing. As was typical for Milwaukee, they built the Titan a little bigger than the Mogul.

A Milwaukee Mystery Tractor?

An undated photo exists in the IH archives of what may be a very advanced IH tractor for its time. Although no date exists on the photograph, a note seems to tie it to Milwaukee Works. Specifications for the tractor are included. This tractor may have been a Milwaukee Works attempt to design a modern tractor in the late teens and may be an ancestor to the IH 15-30 gear drive.

Chicago Works Light Tractors
Mogul 12-25

The first tractor that IH aimed at this smaller market was the Mogul 12-25. Authorized for production in 1912, only four days after the first Mogul 10-20, in many ways the 12-25 covered new ground. Instead of open construction with a canopy, Tractor Works engineers gave it a sheet metal hood. Instead of chain steering, automobile steering was used. Engine rpm were much higher than previous IH practice to get more horsepower out of the smaller engine. The small size and lower cost of the tractor meant more farmers could afford to use it in situations that larger tractors could not handle economically. The first 12-25 was an attractive tractor with exhausts from each side of the opposed engine piped into a single muffler. The sheet metal hood was a little more attractive than the production version.

Although the Mogul 12-25 was an advanced product, International had old stock to clear out. As a result, the Mogul was held back in Canada but promoted heavily domestically. One letter found in IH's advertising archives states the advertising matter for the new Mogul 12-25 must not be sent to Canada, as material sent to the blockhouses (district warehouses in the days when territories were referred to as "blocks") would likely result in some orders even if no information was sent directly to the public. At that time there was a large number of IH tractors built up that had to be sold, and it was felt that the sale of Mogul 12-25s would greatly interfere with the sale of these older tractors. The second letter went to dealers in the United States. This letter implored blocks and dealers to take special notice of these tractors and advertising materials and to push sales of this machine.

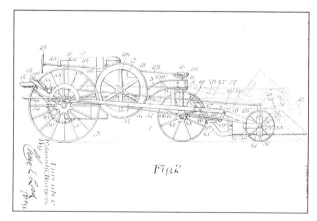

Burgess strikes again, this time with a scraper. Controlling these implements from the operator's seat before hydraulics must have been an adventure. No photographs of the three unusual Mogul applications are known to exist. *U.S. Patent*

The first Mogul 8-16 four cylinder, which later became the IH 8-16. Tiny fenders, a very light frame, and a general look of simplicity mark this tractor, all of which would disappear in production versions. The long strings of boxcars in the background lead to McCormick Works, which loaded and unloaded hundreds of them a day. *State Historical Society of Wisconsin*

Mogul Four-Wheel-Drive

The Mogul 12-25 was also the test bed for what is believed to be International Harvester's first experiment into four-wheel-drive tractors. Designed by Joe Kane, the four-wheel drive photos reveal what looks like a 12-25 with a driven front axle and larger front wheels. Although only one poor-quality photograph exists of the four-wheel drive, the power transmission to the wheels appears to be friction. Not much else is known of the tractor except that some testing was done in Kansas.

Mogul 8-16

The Mogul 8-16 first appeared in the crucial year of 1914, a time when a farm recession was in swing and International's sales of large tractors, along with everyone else's, were down. Harvester's tractor shipments slipped from 3,831 in 1912 to 1,930 in 1913, of which 698 were Mogul 30-60s and 436 were Mogul 15-30s. Of those produced in 1913, 117 were Mogul 12-25s, 120 were Titan 18-35s. And 125 two-speed Type B gear drive models were produced that year.

International had a plethora of tractor models in the line in 1913, with a total of 16 available. Production for 11 of these models was under 100 tractors apiece, with 6 models under 40 tractors apiece. Clearly, IH's tractor business was a little sick.

The tractor business did not improve in 1914. Seventeen models were in production, but production dropped to 1,095 tractors. That year only two tractors were produced in more than 100 copies: the Titan Type D 45 horsepower and the Mogul 12-25.

Several models would be dropped after this disaster, but one bright spot was on the horizon: the new Mogul 8-16. Designed in 1914, the single-cylinder, single-speed Mogul 8-16 was tested and built without much delay. Early Mogul 8-16s had an angle iron and sheet metal water hopper, which was changed to cast gray iron May 6, 1915.

Although produced in only 20 copies for 1914 (probably preproduction), in 1915 8-16 production skyrocketed to 5,111 copies. Higher farm prices as a result of the start of World War I in Europe almost certainly played a part in the recovery, but no tractor in IH production had seen a year of more than 2,000 produced, let alone a year of over 5,000. Given that IH's total production of tractors for 1915 was only 5,841, the impact of the Mogul 8-16 must have been massive on the IH organization and a real feather in the hat of the Mogul people.

Mogul 5-10

Although there was definitely a prototype of this tractor built, it was never released for production or sale, and no identifiable photographs have been found. One reference to this tractor refers to it being a duplicate of the Mogul 8-16 but built to a smaller scale.

Mogul 10-20

The Mogul 10-20 was basically an 8-16 with an increased bore cylinder. Other improvements included a two-speed transmission. The four extra engine horsepower had a wonderful effect on sales. In 1916, the first year of production, the Mogul 10-20 only sold 25 copies (probably preproduction tractors) after production was authorized on November 23. Interestingly, the Mogul 10-20 was authorized over a month after the much more advanced IH 8-16 four-cylinder, indicating either the need for a 10-20 horsepower size tractor in the Mogul line (probably to compete with the Titan 10-20) or some misgivings about the high-speed four-cylinder tractor (the IH 8-16 had a motor governed for 1,000 rpm while the Mogul 10-20 was governed at 400). In 1917, production climbed to 5,338 as opposed to the Mogul 8-16's 665 (the 10-20 probably replaced the 8-16 in production at some time during the year). The Mogul 10-20 was sold until 1919.

International 8-16

From a developmental standpoint, the IH 8-16 was one of the most important tractors for the company. A product of Chicago Tractor Works, the 8-16 embodied many improvements when first built, and would see many more during its production run and continuous improvement by IH's new Gas Power Engineering Department. The original design for the IH 8-16 was done at the same time (1914) as the design for the Mogul 8-16, which may have been actually in competition. Power for the first (and as it turned out, later) four-cylinder 8-16 was a Model G truck engine, which itself did not enter into production until 1916, suggesting that there may have been problems with this particular engine. For whatever reason, the Mogul one-cylinder 8-16 proceeded into production with the four-cylinder 8-16 design getting shelved after testing. In 1916, the IH 8-16 design was revived, with production being authorized on October 16. Compared with the other IH trac-

The four-wheel-drive 8-16 next to the new addition at Chicago Tractor Works. The simple controls of the tractor can be seen clearly. These tractors proved to be very popular in the hands of the few farmers that got them, but were expensive for IH to manufacture. They also had problems crossing drainage ditches. Still, if IH could have resolved a patent problem involving the differential, they might have been produced. *State Historical Society of Wisconsin*

The six-wheel drive showing off the reason for the two extra wheels—the crossing of ditches. However, the extra wheels also made the tractor harder to steer, a real problem in fields without adequate room at the end of the rows. Several of these tractors were produced and tested on California fruit farms, but expense and patent problems again prevented production. *State Historical Society of Wisconsin*

tors in production, the IH 8-16 was years ahead in layout and design. It was a small, compact, and relatively streamlined tractor when compared with other IH products. The size and shape meant that it could be used in orchards and industry as well as the average farm. The tractor was redesigned with a new 4x5-inch motor and, after testing, it was declared to be ready for production. It should have been a good competitor in the fight against the Fordson, but the Titan 10-20 would see most of the action. Bert Benjamin was apparently active in the development of this tractor, probably gaining most of his early tractor expertise here before going on to the Motor Cultivator and Farmall development.

Originally a Mogul designed by Tractor Works, the corporate and legal changes sweeping IH as a result of the government's antitrust case saw the four-cylinder 8-16 produced as an International. At the time of introduction, the International 8-16 received heavy publicity as the tractor of the future. Yet in the battle against the Fordson, it would play second fiddle to the Titan 10-20 while going through three major redesigns of the engine. Tractor Works product change decisions indicate that the

The full crawler version of the IH 8-16. Although traction and flotation would have been good, the manufacturing expense was too high and IH was still trying to learn how to make durable crawler tracks, rollers, bearings, and all those other parts they had never manufactured before! The controls appear to be similar to the four-wheel drive and six-wheel drive tractors. *State Historical Society of Wisconsin*

first production series of International 8-16s may have had serious reliability problems. One letter describes the removal of a tube running from the crankcase to the drive chains. This tube conveyed oil from the engine to the chains for lubrication, but also apparently allowed fresh air into the crankcase, leading to ignition of fumes and explosion of the crankcase. Another major problem regarding the lubrication system in the early 8-16s in the end was solved only with the modification of the Model G truck engine into a replacement engine, both in production and as a replacement for those engines in the field that could not be repaired. The redesign also saw changes made to the radiator.

Four-Wheel-Drive 8-16

Nonetheless, the 8-16 served as a test bed while IH's new Gas Power Engineering Department pondered the future of tractor design. The designers tested 8-16s with several different wheel configurations. The first tried was the four-wheel drive, which was first experimented with back in 1913. In 1917, four-wheel-drive trucks were being produced in large quantities for the Allied war effort in World War I, mostly in Wisconsin factories. The Jeffries Quad (later Nash) was being produced in Kenosha, while the FWD Company was producing four-wheel drives in Clintonville. Other manufacturers were starting to produce four-wheel-drive trucks (notably

A half-track version of the 8-16. IH's histories do not mention this tractor, and the half-track may have been an attachment sold by another manufacturer. The photo is not from the engineering department. This photo was taken June 27, 1919. *State Historical Society of Wisconsin*

Oshkosh Truck, with an engineer from FWD) and even four-wheel-drive tractors (including a product from Clintonville with FWD involvement). In view of all this activity, it is natural that IH examined the possibilities.

Several pictures exist of IH's second four-wheel-drive experiment. Even more interesting are some comments that L. B. Sperry, who became IH's chief of engineering, made about the early tractors in 1944. Sperry stated that although the four-wheeler performed well in the field, it encountered serious problems crossing ditches and other obstructions. Several of these tractors were placed with farmers for testing. (The farmers actually bought them from IH at this time from selected branches. The tractors would be followed as the farmer used them.) A farmer who received one of the test tractors refused to give the tractor back to IH due to its excellent performance, causing severe problems in supplying the man with parts. The emphasis that Sperry gave these difficulties hints that this incident was responsible for

IH's later policy of destroying experimental tractors whenever possible. Almost as much trouble was encountered in tracking them down. Sperry also stated the reasons that the four-wheel drive was not produced was the expense of manufacturing and patenting the differential steering of the tractors that were held by other parties.

In order to correct the problems of the four-wheel drive, IH went to a six-wheel drive. This tractor performed excellently, doing everything asked of it except to turn tightly around at the end of the row. Five of these tractors were produced and sent to citrus groves in California, a state that became increasingly important to IH. However, one problem could not be overcome. The tractor would have violated patents owned by someone who was apparently unwilling to deal with IH. Given the cooperation that existed on patents between the major manufacturers, this indicates that a small or non-farm equipment company owned these patents (or possibly an inventor who thought he should become a mul-

The Six-Wheel-Drive Tractor

International patented the six-wheel-drive tractor, but the patent hints that it wasn't an easy process. The patent that was eventually granted was actually a division of the original application, which was never granted. The original application was applied for August 25, 1917, and was divided and a new application was filed November 11, 1922. The patent, number 1,606,707, was finally allowed on November 9, 1926, nearly 10 years after the original was filed.

The purpose of the patent reveals why International was interested in the layout of the tractor. "The ordinary three or four wheel tractor having one or two driving wheels meets the general needs in connection with plowing and other draft loads, but there are conditions which cannot be combated successfully with this type of tractor. For instance, it is not uncommon to see the two rear drive wheels of a two wheel drive tractor bury themselves up to their hubs when driving in sandy soil or in wet, soggy ground. To overcome this particular difficulty occasioned in certain sections of this and other countries, track laying types of tractors have been built and used. At best, however, the track laying type of tractor is a complicated piece of machinery having a large number of intricate parts which thus far have given considerable trouble to their operators. There is a need, therefore, for a simple general-purpose tractor which also is capable of effective work, especially in sandy, loose, or muddy ground." So reads patent 1,606,707, applied for by Edward A. Johnston and Gustaf W. Engstrom.

The drawings show a tractor obviously derived from the IH 8-16, but with several changes. The wheels were driven through a system of gears (38 is the wheel driving gear, while gear 41 is the power transmitter) running in long casings (21 and 22) located under the main frame. Arms (13 and 14) extend down from the main frame. The arms had long slots that the casings rode in, keeping the casings in line while allowing the casing to pivot up and down. The casings pivoted on trunions.

To drive the tractor, the operator needed to handle several levers (the tractor had no steering wheel). To shift, the operator moved gearshift levers 84 and 91. Gearshift lever 84 handled low speed and second speed only. Gearshift lever 91 handled high gear and reverse. An interlocking mechanism prevented the operator from engaging two gears at once.

Steering was accomplished with lever 98. The six-wheel tractor used a differential brake system to steer. A lever was connected to the brakes. If the lever was moved in one direction, the brakes on that side of the differential would grab, slowing down the gears and wheel on that side while speeding up the gears and wheels on the other side. The harder and farther the lever was moved, the tighter the tractor turned.

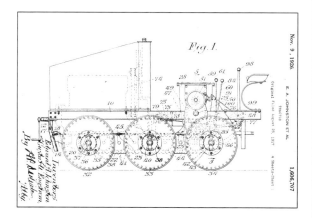

A side view of the 8-16 six-wheel drive. The dotted line outlines of the gear system used to transmit power, labeled 41, can be clearly seen. The enormous amount of gear cutting necessary to produce this tractor was probably a major factor in the high cost of manufacturing that resulted in the tractor not being produced. *U.S. Patent*

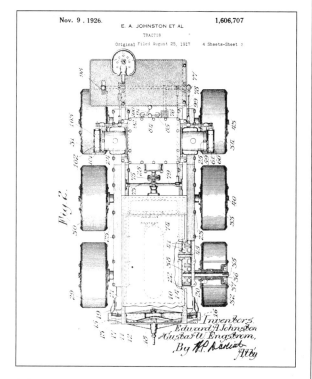

Left
The top view of the 8-16 six-wheel drive shows the casings that the drive gears were located in running along the side of the tractor. The gears can be seen in the cutaway near the lower front wheel. *U.S. Patent*

The Gougis tractor in France. Although this photograph does not show the PTO shaft between the tractor and the binder, it does show a very attentive and dapper Ed Johnston watching the PTO work. Over a decade later, Johnston would resurrect the idea he was watching in action here, and the American farm tractor would never again be the same. *State Historical Society of Wisconsin*

timillionaire). This patent problem killed what could have been a major technological advance for the American farmer.

International 8-16 Crawler

The other configuration experimented with wasn't about wheels at all. The 8-16 saw IH's first major investigation into crawler tractors. Holt, Best, Yuba, and several companies had been building crawler tractors for decades and had made the transition to gas-powered tractors in about 1910. Most of these companies were based in California, which IH was becoming more and more interested in (although crawlers would be of use everywhere and there is no record of IH's motivation in pursuing crawlers at this date). Sperry stated that expense and short life of the track doomed the 8-16 tracklayer at that time. There is a photo of an 8-16 halftrack, with tracks on the rear and wheels in the front. It is not known whether this tractor was an IH experimental tractor or if the tracks were an early accessory produced by another company. Since the picture looks as if it is from Hinsdale Farm, the tractor is more likely an IH experiment.

The History of the Power Take-Off, Part II

International Harvester is often associated with marketing the first power take-off (PTO) in the tractor industry. Although this is true, the development of this attachment is far more complicated. IH had experimented with power take-off with the Automowers, as mentioned previously. Development did not stop there. In 1904 or 1905, French agricultural workers had gone on strike in the Brie region. As a result, much grain was down or lodged, resulting in a mess that was difficult for a reaper or binder to handle. Albert Gougis, a French implement manufacturer from Auneau, France, designed a tractor that had a shaft running from the tractor to the binder. The shaft had two universal joints and was a live PTO. Gougis specifically mentions the tractor could be thrown in and out of gear while the PTO was still running at a constant speed and that the tractor could be driven with the PTO disengaged. Photos reveal this PTO is virtually the same idea as later manufactured by IH. At Gougis' invitation, IH officials, including Ed Johnston, examined and were impressed by the tractor and PTO system. The binder Gougis modified was a McCormick.

Gougis almost certainly saw the Automowers at the Paris World's Fair in 1900. Unfortunately for Gougis, another inventor devised a pickup for the traditional harvesting equipment that accomplished the job of handling the down grain, at an expense of just a few francs, which was far cheaper than buying a new tractor.

The PTO idea languished for 10 years, only to be reborn in several different ways at IH. In 1917, Bert Benjamin watched hemp harvesters at Nappanee, Indiana. Hemp was, of course, valuable at that time due to World War I and the resulting need for rope and fiber. Hemp is a notoriously difficult crop to harvest. The fibrous nature of the stem makes it difficult to cut, while the large size and weight of the plant pose special handling requirements. Harvester built and sold hemp equipment adapted to these special needs until hemp was outlawed in 1937 and again during World War II.

While reporting to the experimental department at the general office about his experiences in Nappanee, Benjamin reported the hemp machines had auxiliary engines mounted on them. The leaves and dust created when the hemp was harvested interfered with these engines, resulting in a 30 percent loss of time from trying to repair the engines. The tractors, operating in front of the binder and out of the dust cloud, were not affected and kept running well. The extra weight of the engine and tongue truck also made the binder sink into the very soft ground at Nappanee. Removing that weight by using the tractor (which could use extensions for extra flotation) would remove that problem. Benjamin went on to say this same idea applied grain binders as well. Benjamin also stated the owner of the hemp farm he had visited could afford three more Titan 10-20s if they were outfitted with the PTO. Benjamin himself believed a PTO

IH built at least 50 IH 8-16s with PTO shafts. IH went through massive expense and efforts to design implements that could take advantage of the new system, re-engineering corn and grain binders and mowers, as well as entirely new implements such as mechanized road graders. The PTO experimentation was successful enough to force a redesign of the 8-16's replacement to include the system. *State Historical Society of Wisconsin*

Although IH didn't attach a caption on this photo, this is probably the first 15-30 built without a cab. By 1918, Ford was sweeping the tractor market, and any way costs could be reduced by IH to compete was fair game. Almost every manufacturer eliminated canopies during these years. *State Historical Society of Wisconsin*

would be a good selling point for farmers and would reduce expenses to IH in regards to the poorly running auxiliary engines. Benjamin urged a trial PTO be made immediately and sent to Nappanee for testing. The records do not show this happened.

In 1918, IH engineers experimented with a PTO to run a mowing machine for the Motor Cultivator. This may have been very similar to the Automower style PTO, which IH didn't really abandon until the 1920s. Later in 1919, IH returned to the Gougis style of PTO for a special binder being built at McCormick Works. These PTOs were designed for the International 8-16. Fifty 8-16s with PTOs were produced, with authorization granted by Tractor Works decision number 451, dated March 29, 1919, and described by a drawing dated April 2, 1919. These were apparently the first production PTOs built by any company. After testing was successful, IH offered for sale the "IH Tractor Binder," which was actually an 8-16 with a PTO plus a McCormick grain binder equipped with the necessary shaft and transmission to transmit the power to the various parts of the binder.

Photos reveal several ideas competing within IH in the period of 1919-1921 regarding the structure of PTO-operated equipment. Deering Works proceeded along two engineers' paths about tractor mowers, while McCormick Works also produced its own ideas about what a tractor mower should look like. Positions of the mowers included under the tractor and at the rear of the tractor, with McCormick Works experimenting with two mowers in tandem behind the tractor.

Very little information exists about the 12-22, other than a few photos and some specifications. The tractor featured a four-cylinder engine laid on its side, with the starting being done from the driver's seat. The tractor was apparently designed and built at Milwaukee Works and may have been an early attempt to build a tractor to fill the role of the McCormick-Deering 15-30. Unfortunately, the photos and specifications were not dated. *State Historical Society of Wisconsin*

Another path of experimentation involved what could best be described as a powered disk plow, although this is not entirely accurate. This implement was an orchard tool powered by a PTO to reach and cultivate under trees. What is interesting about this tool is that the side reach required fostered a second type of PTO that is forgotten today. While rear-mounted PTOs are common and front-mounted PTOs are also used (IH engineers actually experimented with the front PTOs for water pumps and an early version of the Electrall during the McCormick-Deering 10-20 years), the early experimenters also tried side-mount PTOs. Side PTOs attached to the belt pulley shaft. What would be unusual ideas today were equal possibilities then when the PTO was a new idea.

The Last Ohio Tractors

By 1915, the Ohio Manufacturing Company knew that a new tractor design was needed. A new design was made for a much smaller tractor of light and compact design, the Whitney 6-12 (Whitney was president of Ohio Manufacturing, which he later renamed after himself). The new tractor was attractive and capable. An opposed engine (probably a Doman or Giles) provided power. International Harvester marketed these tractors for a short time and published one piece of advertising about them in 1915. The Whitney tractor was probably made unnecessary by the success of the smaller Moguls, although Whitney would market larger sizes of the same design until the early 1920s.

The Whitney 6-12 in the form it was sold in by IH. The Whitney was named after A. B. Whitney, the president of Ohio Tractor Company. After building hundreds of tractors for IH, Whitney needed a new, more modern design. The engine was probably a Doman or Giles. The 6-12 was built by Ohio Tractor, although it is far from clear whether IH was the only sales agent. Larger versions were sold by Whitney in later years. *State Historical Society of Wisconsin*

Chapter 4

Development of the McCormick-Deering Tractors

When 1917 began, the world was a much different place than the previous years. The tractor business was well established, with several competitors making and selling good tractors. World War I and its resulting manpower, horse, and food shortages had created an ideal marketing situation for the tractor. The Bull tractor had come and was now nearly gone, leaving a legacy of small tractors. The biggest tractor that IH built was the Titan 15-30 (except for a very few Titan 30-60s). The main production would be of Titan 10-20s in Milwaukee, with the International 8-16 produced in Tractor Works and Plano Works. The Titan 15-30 would be produced in Chicago after being relocated from Milwaukee. Mogul 10-20 production would continue for a while, but would be discontinued by 1919.

International was making small tractors with a good reputation at a rate that it could generate profits. The implement lines were doing well. Stationary engine production would soon be concentrated into the Type M line, eliminating a large number of different lines of engines. Larger stationary horsepower would be handled by power units derived from truck, harvester-thresher, and tractor engines.

Then things started to go downhill.

The government finally issued a judgment in the antitrust case dating from 1913. The old Milwaukee, Osbourne, and Champion that helped make the original International Harvester Company lines were ordered sold, leaving Harvester with McCormick and Deering. In addition, IH could only have one dealer per city or village. While divesting itself of the old line implements of Milwaukee and Champion would eliminate duplication without reducing sales by much (indeed, it was hard to find a buyer for either line), the dealership ruling meant that there were plenty of dealers out there looking for a new supplier.

Another major issue facing Harvester was the future of gasoline supplies. Harvester had been selling kerosene tractors for several years, claiming that its tractor engines were true kerosene engines rather than gasoline engines which could burn kerosene. Supplies of gasoline had become even more in demand and higher priced. At that time, only natural gasoline was used (included in the oil as pumped from the ground). The success of the automobile and truck (and to a more limited extent, airplane and tractor) meant that gasoline was a product in

IH produced two different versions of the experimental 15-30. This is the channel frame version, which had two parallel steel channels for a frame, much the same as all earlier IH tractors. Although this tractor performed well, the unit frame (cast frame) 15-30 protected the working parts better, which was one of IH's main goals for the new tractors. This tractor is seen at Chicago Tractor Works in January 1920. *State Historical Society of Wisconsin*

The unit frame 15-30 won the competition with the channel frame 15-30, putting in impressive amounts of work without stopping for maintenance. Hood louvers are seen on the left side of this 15-30, although not very many. This tractor is similar in appearance to the drawings appearing in the patent that covered the basic 15-30-10/20 design. *State Historical Society of Wisconsin*

Another view of the 15-30 in 1920. IH was at this time examining several different ideas in tractor construction. After this photograph, the tractor was probably shipped out for testing in the field. If shipped in January, the tractor almost certainly would have been shipped to California (where many retired IH executives lived and kept track of the company's activities) or Texas, where IH was only starting to do business again after being banned from the state. *State Historical Society of Wisconsin*

demand, a far cry from its previous status as a dangerous waste product. Commentators predicted a catastrophe if substitutes were not found and popularized. While the ultimate answer was diesel and catalytic cracking (a process which produced more gasoline from the oil than naturally existed by breaking the crude oil into its components and rearranging them into different petroleum products), kerosene was the first obvious path. In fact, in the later teens, IH sent regular bulletins to its tractor and engine salesmen comparing gasoline and kerosene prices for dozens of locations around the country.

The basic unreliability of the currently existing farm tractor was another problem. The construction of the tractor was open, letting dust into gears, transmissions, engines, and other critical areas. Dust rapidly decreased tractor life. An engineer from a different company noted one location in the United States where tractors lasted an average of about one year due to fine powdery soil that quickly entered and destroyed unprotected bearings. An anonymous farmer noted that his tractor had cost him $1,000, but that the manufacturing company had already allowed him $400 in free repairs in the first year alone. Many executives at Harvester were unhappy with tractor performance and the value that the farmers received from their tractors.

Another view of a 15-30 experimental tractor during January 1920. Notice the lack of any provision for a PTO shaft. This photo is taken in what was apparently the storage yard of the gas power engineering department at Chicago Tractor Works. The tractor marked 11 on the front is an IH 8-16 four-wheel drive and appears in photographs for years. *State Historical Society of Wisconsin*

Another, different unit frame 15-30. This particular tractor is without louvers in the right side curtain, and this is the only photograph in which this is seen. This tractor is probably the next level of development from the original 15-30. Again, probably a January 1920 photo. *State Historical Society of Wisconsin*

The 15-30 shortly before entering limited production at Milwaukee Works. The side curtains have been changed, while a small fuel tank similar to that of the 10-20 is seen. The photograph is probably from spring 1921. *State Historical Society of Wisconsin*

The 15-30 as it entered production in 1921. This is tractor number TG-135, owned by W. D. Stocking of Lindenwood, Illinois. In an affidavit dated April 16, 1935, Stocking stated that he had bought the tractor in 1921 and had used it ever since. Only a few worn parts had been replaced. He paid a total of $1,351.68 for the tractor, including freight. IH 15-30 production would remain low until 1923, when the Titan 10-20 ended production and IH got an assembly line into Milwaukee Works for the 15-30. *State Historical Society of Wisconsin*

IH's 8-16 chain-drive tractor encountered a large amount of problems in farmers' hands. Already by 1921 IH was testing a replacement. Based on the larger 15-30 tractor (at this time rated 12-25), the 8-16 would not see production for another two years as a result of having to re-equip Chicago Tractor Works with an assembly line and a redesign with a PTO shaft into the 10-20. This tractor is seen February 1, 1921. *State Historical Society of Wisconsin*

IH started engineering an industrial version of the 10-20 gear drive before the agricultural version started large-scale production. Here we see an experimental industrial version in a photo taken December 1, 1922. The disk wheels are unusual. IH tested the early industrials in their own factories, where among other tasks, they helped move the machinery to set up the assembly lines for the tractors. *State Historical Society of Wisconsin*

This 10-20 industrial is one of the first IH experiments with pneumatic tires. Industrial tractors received pneumatic tires well before their agricultural cousins, mainly because of the hard surfaces they operated on. Other reasons for the need for pneumatic tires were increased traction and the greater amount of wear that industrial tractor wheels encountered, again due to hard surfaces. *State Historical Society of Wisconsin*

IH experimented with a dedicated tractor for pulling trailers at higher speeds. The tractor was apparently based on the 15-30. This example was photographed on December 2, 1925. This idea was tested in California, where photographs reveal that this tractor (or another like it—production numbers are unknown) could pull a big load under the palm trees. The 15-30s were used for freighting in Great Britain, where they received similar cabs but did not receive (so far as is known) the chassis modifications shown here.

The first 15-30 industrial versions were basically agricultural versions with rubber tires and were marketed that way. Shown here is what was probably the first of these tractors. IH sold these tractors as 15-30s with industrial wheels, then sold the Model 30 industrial tractors for several years, which had serial numbers separate from the regular 15-30. After the introduction of the I-30, the Model 30 was discontinued and 15-30s with industrial wheels were again sold. *State Historical Society of Wisconsin*

Perhaps the biggest problem of all was Henry Ford. Ford's first introduction to power machinery was a steam tractor which he saw coming down a road near his home as a youth. By 1917, Ford had experimented with tractors for several years, but the coming of the war had presented new opportunities. Ford proposed a plan to the Allied governments for production of small, cheap tractors that would meet the needs of small farmers, while allowing for standardization so that parts could be readily supplied. While production was at first earmarked for England and Canada, Ford clearly had his eye on the United States market. Ford started a public relations campaign reaching into the homes of nearly all Americans. Many magazines and newspapers proclaimed Ford to be the inventor of the tractor, a claim that the rest of the industry bitterly contested.

International began development of a new 15-30 (known as the 12-25 during certain times) tractor in 1917. This project soon had the goal of meeting all the concerns listed above. The tractor had to be powerful, cost-effective to build, buy, and use, long-lived, and able to put the Fordson in its place.

For the first time, the old system of each works having its own design team was abandoned. Instead, the engineers were gathered at Chicago into the new Gas Power Engineering Department. Engineers came from Milwaukee, Tractor Works, and other locations to Chicago to engineer tractors, as well as gas engines, some trucks, some other experimental machines (eventually including heating, cooling, refrigeration, a steam locomotive, and all sorts of other goodies), and manufacturing processes. For the first time, IH's top tractor engineers and experimenters were working together. The design team included Ed Johnston (from Tractor Works), L. B. Sperry (from Milwaukee Works), D. B. Baker (from Tractor Works), and dozens of others. While each works retained an engineering staff, the main design work was concentrated at Chicago Tractor Works.

Two prototypes were produced in 1918. One had a frame made out of channel steel; the other had an integral frame that would later enter production. Both tractors were tested extensively in the marshes around Kankakee, Illinois, probably the most difficult testing area available within close proximity to the engineers in Chicago. Although both tractors performed well, the integral framed tractor was chosen for production. This tractor continued to be tested, putting in performances that shocked onlookers used to the then-current tractors.

It should be noted that International called its integral frame a unit frame in most documentation. Other companies and publications have referred to this kind of frame as a "frameless" type. Whatever you call it, this type of frame is essentially a bathtub-shaped piece of metal. The engine, transmission, and final drive are nestled into the inside of the tub, with

A very early orchard conversion of the 15-30, the sheet metal was originally designed and applied in California, either by an IH employee or a farmer. The design was then copied in Chicago, which produced this version. It is not known whether IH produced this version or not. *State Historical Society of Wisconsin*

the wheels and running gear bolted to the outside. Notable examples of tractors that used this construction are the Wallis Bear, which is credited as the first integral frame tractor, and the Fordson. For the purposes of this book, this design will be called an integral frame, be it an International design or otherwise.

The first 15-30s were produced in 1921, probably on a preproduction basis, production and sales beginning in earnest in 1923. It took at least two years for IH to outfit the Milwaukee plant with the necessary equipment. IH made a massive investment in assembly lines and tooling just as the early 1920s farm depression hit. When combined with the price war with Ford, IH must have tied a large amount of its available capital into these two tractors.

At some time after the designing of the first 15-30s,

A more normal orchard tractor, showing the typical wheel fenders. On this tractor, the steering wheel has been lowered by offsetting it to the right and by using the belt pulley location to base it. California orchard versions were shipped with the belt pulley disassembled, indicating that they used this configuration.

A 10-20 orchard version, with wheel fender aprons and lowered steering wheel. This photograph was taken May 7, 1924. The 10-20 orchards were far more common than the 15-30s, for a variety of reasons, including price and size. There were numerous versions of the 10-20 orchard produced, differing mainly in transmissions and front axles, as well as special equipment for certain localities such as the Jacksonville, Florida, branch with its specialized citrus crops.

This 10-20 has unusual shields that are not seen in other photographs in the IH collection. This tractor was probably intended for the nursery market, where young trees and shrubbery had low-hanging, delicate branches that needed to be gently brushed away from the wheels. This photo was taken May 7, 1924. *State Historical Society of Wisconsin*

engineering of a smaller companion to the 15-30 began. The 10-20 followed closely in the footsteps of the 15-30, with similar construction throughout the tractor. The major outward differences in appearance (other than size) were the smaller gas tank and fewer hood louvers. In a surviving line drawing of the "gear drive" tractors, the same drawing is used for the 10-20 and the 15-30, but with different dimensions.

Interestingly enough, although the McCormick-Deering 10-20 obviously replaced the Titan 10-20 in both size and popularity, the McCormick-Deering 10-20 was originally rated at 8-16. The original push for authorization of this tractor was made in October 1920, and failed, indicating the tractor was not yet ready. The

Orchard 10-20s were offered with disk wheels for most of the 10-20's existence. The purpose of disk wheels was to prevent branches from getting entangled with the wheel spokes. This photograph was taken February 25, 1925. *State Historical Society of Wisconsin*

The need for a narrow-tread 10-20 is less obvious than for a narrow-tread Farmall. It did sell 1,400 copies over eight years, probably for orchard use and other specialty crops. Narrow-tread tractors seem to have been controversial within IH, with the decision to manufacture being approved and canceled several times. This is among the first photos of the 10-20 narrow tread, taken June 10, 1924. *State Historical Society of Wisconsin*

This is one of only a few photos that appear of the 15-30 wide-tread tractor. The purpose of the wide-tread modification is unknown, but probably has to do with someone's need for a more stable tractor, or perhaps the need to mount some special piece of equipment. The tractor (or modification) serial number is Q-732. Date of photograph is February 14, 1929. *State Historical Society of Wisconsin*

A 10-20 narrow-tread orchard tractor with another rare or unique modification—a shorter wheelbase. The front axle on this tractor has been relocated farther under the tractor than normal. Orchards often had tight corners in which a shorter tractor could have been an advantage. *State Historical Society of Wisconsin*

tractor was changed, gaining 200 pounds and 2 inches in length. Turning radius went from 27 to 30 feet, and the transmission gearing was changed. Another major change was the jump to a 10-20 rating, but the bore, stroke, and rpm were not changed. Although production was authorized at this horsepower and design on August 25, 1921, production did not take place until 1923 when the assembly lines were ready, and yet another change decision had authorized production (dated December 29, 1922, with different transmission gearing, again). The 10-20 was to see much modification throughout its lifetime, more than even the International 8-16, into different configurations and uses.

Industrial Models

International had entered its farm tractors into the road-building market since the beginning. In the early years, the road market was almost the same as the farm market, as farmers were responsible for maintaining roads bordering on their farms in some places. Road maintenance and building was also a source of income for many farmers in seasons where there was spare time away from the farm. In the early days, even the equipment wasn't that much different, as in the case of "road plows," a super heavy-duty plow that could stand up to plowing the compacted dirt roads. Other equipment, such as rock crushers and asphalt machines, relied on belt power.

By the 1920s, specialization and new opportunities had shifted demand from the large tractors to the smaller tractors even in industrial applications. Rubber tires for hard surfaces were required. Heavy-duty road wheels were demanded. Small tractors had opened up a whole new field of possibilities in facto-

A 10-20 with an electric PTO. Of course, mounting the generator of the front of the crankshaft enabled the PTO to be run whether the tractor was in gear or not, allowing it also to be turned off independently, an important advantage that would take IH another couple of decades to put into production. Electricity in farm implements would also be a subject that IH returned to every few years. The serial number of the modification (it is unknown whether it applied to the generator or tractor) was Q-911. Weight was 4,650 pounds. Date of photo February 13, 1930. *State Historical Society of Wisconsin*

A 10-20 with a side-mounted PTO. During the early years of the PTO, IH tried nearly every possible configuration, learning over time what worked and what didn't. It is unknown what the implement was for, but a likely guess is it was an orchard tool. Given the configuration, it could easily reach in under trees. Of course, terracing and road maintenance are other possibilities. Photo dated February 4, 1925. *State Historical Society of Wisconsin*

This photo gives a good close view of the components of the early IH steam tractors. The complexity of such a tractor had to have been a reason against production. You can see the condenser at upper left, the water and fuel pumps at center bottom, the fuel tank behind the belt, and the boiler (wrapped in asbestos insulation) under the exhaust hood. The engine is at upper right. *State Historical Society of Wisconsin*

ries, airports (which themselves were new), municipal service, and delivery. IH's first true industrial tractor was the International 8-16, but IH's first major industrial success would come with the McCormick-Deering 10-20, and to a lesser extent the McCormick-Deering 15-30.

Industrial development started for the 10-20 in 1922, even before production of the farm versions began. There are photos of experimental or preproduction Industrial 10-20s helping to set up the assembly lines for the new tractors. The early versions had only cast iron disc wheels and solid rubber tires as differences from the farm versions, but further development began in 1923, as well as production of the first 23 tractors for sale (a decision authorized March 16, 1923). By 1924 the Industrial 10-20 had a solid mounted lay back seat, foot accelerator, and spring-mounted front axle. Bottom exhausts, dual rear tires, and rear-wheel brakes were also applied. Again a small number was produced.

By 1925, the 10-20 Industrials were improved even further. The lazy back seat was spring mounted, which

The first IH steam tractors looked like this. The photo, taken sometime in 1920, shows a large condenser, a jacketed steam engine and boiler, and a very large (for a tractor) belt that ran the fuel pump and the water/oil separator that was necessary to remove the cylinder oil from the condensed steam before it returned to the boiler. Oil in a flash (or any other) boiler spelled big trouble with foaming and other maladies. *State Historical Society of Wisconsin*

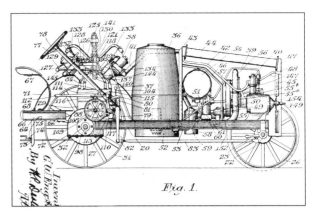

A side view of the Gus Engstrom steam tractor patent. The tractor shown is from fairly late in development. In this view, the equipment inside the sheet metal can be seen. The boiler equipment was a different patent. *U.S. Patent*

was good because the tractor now had a gear good for 10.4 miles per hour. The spring-mounted front now had transverse leaf springs. A rear plate made mounting drawbar hooks easier. IH also designed several attachments for the tractor, including different wheels, lugs, and extensions, and front and rear bumper plates for pushing. Several hundred Industrial 10-20s were produced in 1925, not only for traditional customers, but also for other manufacturers who used the 10-20 as a base for their equipment. The most common use was for one-man-powered graders, but eventually well over 40 manufacturers would base their equipment on the McCormick-Deering 10-20 for uses ranging from mobile cranes to railroad locomotives to dump bodies. IH would also build special versions for certain customers. The "New York City" tractors are perhaps the best example, when IH added a cab and special plow for use in clearing crosswalks and other areas. Narrow-tread industrials

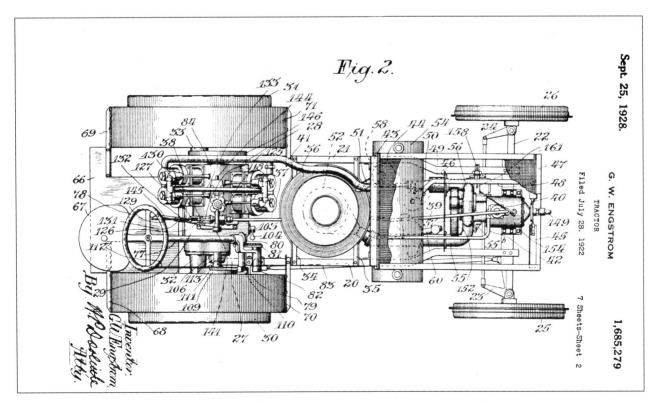

The top view of the Engstrom patent. Here is the four-cylinder V-type engine, and the top view of the fan, auxiliary engine, and pump equipment at the front of the tractor. *U.S. Patent*

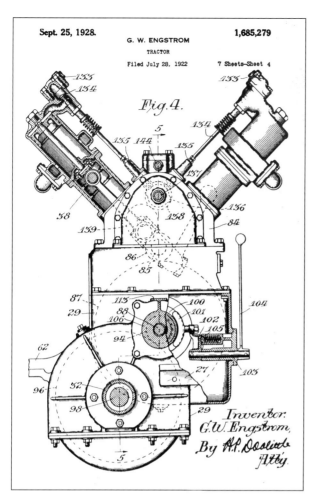

The engine and transmission assembly. Like the 10-20 and 15-30, the engine and transmission of the steam tractor used an all-enclosed, one-piece bolted together construction for greater rigidity and dust protection. *U.S. Patent*

were also built for clearing sidewalks.

The 15-30 industrial versions did not enjoy the same success of the 10-20s, probably because of their larger size, which caused problems in the small alleys and hallways of factories. The first 15-30s for industrial uses were farm 15-30s fitted with industrial wheels. In 1929, an industrial version (Model 30) was built using the same principles of the 10-20 Industrial (now renamed the Model 20). The Model 30 would be only marketed until the I-30 came out in 1931, but apparently some 15-30s with industrial wheels were produced after the Model 30 was replaced.

Orchard Models

Orchard tractors originated in California around 1910. The need to cultivate under trees without damaging branches or leaves demanded a very low, small, shielded tractor. IH first got into the orchard market with sprayers powered by its smaller engines, but had to wait on the tractor market until the Mogul 12-25 entered production. An orchard modification was made,

removing the cab and adding primitive branch deflectors to the front of the tractor. Even with the small Moguls and the Titan 10-20, modifications to the tractors were minimal, usually just orchard lugs for the wheels. With its swept-up hood and small size, the International 8-16 was a natural for the orchard market, but the wheels did not receive shields and other modifications were minimal. The first IH tractors to be modified and sold specifically as orchard tractors by International were the McCormick-Deering 10-20 and 15-30.

The 15-30 was probably the first tractor to be modified. It is known that a 15-30 was modified in California and photographs were sent back to Chicago, where either the original tractor was shipped or a new tractor similar to the original was made. It is not known whether the original modifier was a farmer or an IH employee. Soon, the Gas Power Engineering Department began a series of modifications to 15-30s and 10-20s, including shielding, steering wheel and seat lowering, and PTO equipment. Creativity was clearly running rampant in the mid-1920s with new tractors and new uses.

Model 10-20 orchards were authorized for production October 15, 1923, but much confusion surrounded orchard tractor production, even among IH engineers. Not only did IH have to deal with regular orchards, but it also had to deal with California and its special demands. Several changes were authorized, unauthorized, and then authorized again.

Model 10-20 orchards were built until at least 1938. Some of the final versions built were for Florida citrus groves and were referred to as "Jacksonville" tractors, after the branch they were shipped through. The Jacksonville tractors used a combination of 10-20 parts with the W-30's front axle and steering and an engine improvement package coming from the T-20 TracTracTor. Production of this tractor was authorized August 8, 1936. These tractors apparently worked well, as IH modified the tractor and made it the standard production version (however, only the standard tread version was used). This tractor was a combination of 10-20 and W-30 parts, as well as a T-20 orchard air filter top. The new orchard tractor replaced six 10-20 orchard models and was first produced in 1938.

Narrow-Tread and Wide-Tread Models

The Gas Power Engineering Department was responsible for filling the needs of users other than orchard and industrial users. Such customers were not always happy about where the wheels were placed on the tractors. Unlike other modifications, changing the tread or wheelbase of a tractor is a major operation, involving different axles, transmission cases, and other parts. Engineering came up with several combinations.

Narrow-tread tractors are by far the most known modification. Narrow treads fit situations, such as narrow rows or tight spaces, where the ordinary tractor doesn't fit. Argentina was a notable market for narrow-tread tractors and so were certain regions of the United States. Industrial markets also used the narrow treads, as well as orchard tractors, when tight spaces were encountered. IH produced Narrow-Tread 10-20s for farm, orchard, and industrial purposes.

The first decision to manufacture narrow-tread 10-20s was made January 9, 1925, and then canceled. The engineering department built one tractor to that design, which had a tread of 48 inches. However, the original tread on the decision was 50 inches, which was scratched out. Another decision authorized production at 48 inches on October 16, 1925, but this was changed back to 50 inches by a third decision dated February 4, 1926. This decision was canceled in turn by another (that's four), dated April 17, 1926, changing the tread back to 48 inches. This final decision was canceled, going back to the February 4 decision. There appears to have been a conflict regarding how narrow was narrow enough.

Wide-tread tractors are far less known, although IH experimented with these as well. Wide-tread tractors never caught on in wheel tractors as much as they did in crawler tractors. Wide treads most commonly were used for greater stability on hillsides, making the tractor more difficult to turn over. In addition, some row sizes or crops may have required that the treads be put in certain locations. International experimented with and may have produced a few wide-tread 15-30s and 10-20s.

IH also made at least one short wheelbase version of the 10-20. This sole photograph shows orchard fenders, so apparently there was some need in orchards for a very short tractor. Other details are still unknown.

Alternative PTOs

During this time, International Harvester was still experimenting with power take-offs. Other than the standard rear shaft, front and side PTOs were also tried. The front PTO had the advantage of always being live, powered without having to have the transmission in gear or the main clutch engaged. Both mechanical and electrical systems were tried. The electrical system may have been an attempt to provide lifts for implements as well as power.

The side PTO ran off the belt pulley shaft. One advantage of a side PTO would have been in orchards where low-profile power-driven machinery could be located to the side of the tractor to operate under trees. The tractor would not get under the trees, reducing tree and tractor damage.

Model 22-36

In 1929 production of an increased horsepower version of the 15-30 began. The design work probably took place in 1927. This tractor had an increased cylinder bore and taper bearings instead of straight roller

As years went by, the IH steam tractor acquired more sheet metal and a more polished look. Here, the condenser has been relocated, the pumps and piping have been covered by a sheet metal hood, but the engine is still exposed. A very faded International decal is seen on the hood. This photo is dated August 11, 1922. *State Historical Society of Wisconsin*

bearings. There were several problems with these bearings. At first, many were not properly adjusted when assembled, although there was a claim made that the roller cage did not provide enough clearance for the rollers to roll freely. Other faults were still being dealt with in October 1930. The net result of these problems was vastly increased wear in the transmission. According to an October 1930 letter, tractors dating from 1923 were showing less wear than a two-year-old 22-36.

In one of the last photos of the steam tractor, a good paint job has been applied, the boiler is painted black, and the cylinders of the engine have been enclosed with sheet metal. A smaller lip is seen below the front of the condenser. The attractiveness of the photograph makes one wonder how close IH executives were to producing the tractor—or how desperately the engineers wanted to continue. But the 15-30 and 10-20 had by now entered full production. The photo was taken July 18, 1923. *State Historical Society of Wisconsin*

This view of the 12-25 shows the radiator, which is very similar to the design of the IH 15-30 chain drive. The stamp in the upper right is the gas power engineering department's experimental "received" stamp, dated March 20, 1920. The frame of the 12-25 is definitely similar to the unit frame of the 15-30 (which was engineered to have major components fit inside, rather than having the frame be a part of the components). The radiator looks similar to the design used by the IH (formerly Titan) 15-30 chain drive, while the engine itself looks like a modernized version of the Titan 10-20 engine. By this time all were proven components, and the tractor would have been cheaper to produce than the four-cylinder ball bearing crankshaft 15-30 gear drive. *State Historical Society of Wisconsin*

The 12-25 was being tested at the same time as the 15-30 gear drive (March 1920), leading to the conclusion that it was a competing design. (The 15-30 was rated a 12-25 during this time.) Although little is known of this tractor, a look at the photos suggests that the 12-25 used a combination of features from other IH tractors. *State Historical Society of Wisconsin*

Neither the 10-20 nor the Farmall had been modified to use taper bearings up to that point except for wheel bearings, where no trouble had been experienced. The bearing problems were eliminated by the correction of production problems and some revisions to lubrication, heat treatment of transmission parts, and changes in seals on shafts and other parts passing through the transmission walls.

Model 12-25

The 12-25 was developed as another alternative to the integral framed McCormick-Deering tractors. A few photographs are all that remain of this rare tractor. The 12-25 was photographed in March 1920, apparently before going out for testing in the field. The tractor appears to be a combination of the unit frame, a radiator similar to the IH 15-30 Chain Drive and an engine similar in appearance to the Titan 10-20. No mention of the tractor is made in the various IH histories, indicating that the tractor was not an important program, or was not very successful.

Experimental Steam Tractors

International did not put all of its eggs into the 15-30 and 10-20 baskets. Due to the fuel and longevity questions about gas and kerosene tractors, IH looked into the possibilities of modernized steam tractors. New developments in steam technology made an investigation of the possibilities attractive. Flash boilers, as adopted from automotive practice, were tried. These boilers allowed steam to be raised quickly, while making a smaller boiler possible. Gearing could be enclosed, while a simpler transmission was possible using the throttling ability and reversibility of steam engines. The engine itself was a Uniflow V-4 (which increases efficiency by avoiding used steam, cooling the valve system by having a separate steam exhaust). A condenser was used that increased energy efficiency while almost eliminating water consumption. Boiler pressures ranged from 500 to 600 psi. All parts were completely enclosed and protected from contamination from dust and dirt.

Patents for steam tractor parts are very interesting. Many of the patents revolve around the burner assembly. IH wanted to burn kerosene and other heavy oils. However, the company also wanted a low-maintenance boiler, so soot had to be eliminated through complete burning. The patents show the evolution of burner design. The engineer (C. A. French) placed the burner on top of the boiler so the residual heat from the boiler could atomize the oil. The location helped eliminate the pilot light as well. French eventually discovered how to bring the burner to full

Steam Tractor Technology

Patent 2,685,279 covers the engine and chassis of International's steam tractor. The second, third, and fourth paragraphs of the patent sums up the purpose and result of IH's research: "Steam tractors are easier to operate than tractors propelled by internal combustion engines but heretofore have been available in large sizes not suitable for use on small farms and for general power purposes, since it has generally been considered impractical to build steam tractors in small sizes.

"The object of my invention is, therefore, to provide a steam tractor that may be made in small sizes of equal or greater power and of less first cost and operating expense than the corresponding sizes of gas engine tractors now on the market.

"It is also an object of my engine to arrange and correlate the several parts of a steam propelled tractor so as to provide a neat, compact, and efficient power plant and tractor."

So wrote Gus Engstrom in his patent application of July 28, 1922. Granted on September 25, 1928, the patent describes a steam tractor meeting Engstrom's criteria. The frame was made of steel channels, supported at the front end by the front axle, and at the rear by the transmission and differential casing, very similar to the system used on the F-12 a decade later. The engine, transmission, and differential were all enclosed and directly connected, keeping dust out and maintaining close alignment, just as similar construction benefited the competing 10-20/15-30 design.

The engine itself was a Uniflow engine, meaning that steam would enter the engine through one valve and leave through another. In traditional steam engines, the same valve was used for both intake and exhaust. The problem with this system was that the exhaust steam would cool the valve, and the valve would then cool the intake steam, lowering efficiency.

IH's design, through the use of the Uniflow system, maintained the high efficiency needed to reduce engine size and to increase economy, which was inherently a problem in the external combustion steam engine, which lost much energy through the smokestack. International's Uniflow engine was a V-4. The crankcase (part 85 on the patent drawing) was bolted to the transmission and differential case 84. The engine was throttle governed using a conventional steam governor. However, the method of controlling steam cut-off was very unconventional. Steam engines do not derive all of their power from the pressure of the steam.

One of the physical properties of steam is that it expands. Steam engines derive power from this expansion. To allow the steam time to expand, the supply of steam is cut off at some point during the travel of the piston well before the piston nears the bottom of the stroke. The higher the rpm, the earlier the cut-off. Most steam engines used a sliding link mechanism to control cutoff, but the IH steam engine used multiple cam sleeves that slid on the camshaft. The operator could select which cam was operating the engine, varying the cut-off. The cams could also be arranged to operate the engine in reverse. The engine in the steam tractors had two different cams per intake valve, for a total of eight, but the patent noted that any number or any different cam profile could be used.

force quickly without torches by regulating the amount of fuel supplied to the burner when starting. Eventually IH had a very advanced, quick-starting burner providing complete, blue combustion of heavy fuels. Another engineer, Gus Engstrom, concentrated on the tractor chassis, engine, transmission, and boiler. The transmission for the tractor was very advanced and showed some influence from the 15-30/10-20 program. The engine, transmission, and axles were in one piece, eliminating dust and rigidity problems.

Chapter 5

Motor Cultivator and Farmall Development

One of the indicators of success of the tractor was measured in terms of how many tasks tractor farming was adapted to. One of the most tedious tasks, and one that resisted power farming, was cultivation. Old-timers still speak of long slow hours spent looking at the rear ends of two or more horses, often on the hottest days of the year. Sulky (riding) cultivators eased the situation for the farmer (or often the farmer's children—the author's grandfather started at age six!), but the higher speeds made tractor farming an attractive proposition.

Tractors meant for duties beyond plowing and belt work are referred to as general-purpose tractors. Many of International Harvester's employees had ideas about general-purpose tractors decades before the first Farmall, the tractor ack-nowledged to be the first general-purpose row crop tractor produced in the world. E. A. Johnston, in his patent application for the McCormick Automower, patented using the power for the engine through the PTO for other implements besides the mower.

According to *Notes on the Development of the Farmall Tractor*, J. F. Steward started working on a motor-powered harvesting machine in 1902. He wrote Alexander Legge in September 1910 to discuss a tractor that would pull plows, propel a binder, a stripper (an early form of combine originating in Australia), bull rake and

stacker (hay tools), corn cutters, mowers, corn huskers, and other machines. Steward said, "What we want is a tractor that will most nearly abolish expensive labor on the farm, both of brute and man." His experimental tractors, mentioned in chapter 3, were aimed at the general-purpose markets. H. C. Waite's 1913 tractor, mentioned previously, was aimed at doing all the work on a farm of 200 acres.

One brief fad that passed through tractor circles was electric farming. C. E. Lucke, a consulting engineer, proposed tractors equipped with generators on the ends of the field connected by long extension cords to various farm machines including harvesters and cultivators powered by electric motors. One wonders if Nikoli Tesla's experiments with wireless transmission of electricity were an inspiration here. IH engineers did hook up electric motors to run belt implements to test the idea.

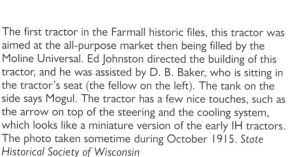

The first tractor in the Farmall historic files, this tractor was aimed at the all-purpose market then being filled by the Moline Universal. Ed Johnston directed the building of this tractor, and he was assisted by D. B. Baker, who is sitting in the tractor's seat (the fellow on the left). The tank on the side says Mogul. The tractor has a few nice touches, such as the arrow on top of the steering and the cooling system, which looks like a miniature version of the early IH tractors. The photo taken sometime during October 1915. *State Historical Society of Wisconsin*

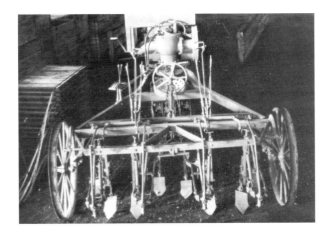

The real start of motor cultivator development occurred in 1916 with the building of this machine. Developed by Carl Mott and Ed Johnston, the Motor Cultivator is shown in the McCormick Works experimental department. This is the same machine as the first Motor Cultivator patent granted to IH. The photo was taken January 5, 1917. *State Historical Society of Wisconsin*

The First General-Purpose Tractor

E. A. Johnston, D. B. Baker, and others designed a Mogul general-purpose tractor, completing the experimental tractor in 1915. The general-purpose tractor was a considerably different beast than the general-purpose tractors of the 1930s. Since farmers had what was considered "huge" investments in horse-drawn farm equipment, it was thought a general-purpose tractor would have to also be able to use that equipment in order to be sold in large numbers. Designs had to accommodate horse hitches, as well as place the tractor operator in position to operate the implement, a hard thing to do since normally the person operating the horse-drawn implement was on top of the implement instead of in front of it. The early general-purpose tractors were never very successful, although the idea itself was good. The Moline Universal was perhaps the best known of the early general-purpose tractors. The Allis-Chalmers 6-12 was designed at the same time as the Mogul, but not put into production until later.

As mentioned, the Mogul was designed to hitch on horse implements, including cultivators. The two surviving photographs show the cultivator retaining its own seat and operator shifting the cultivator gangs. Johnston stated that R. W. Burtis designed attachments for tractors in 1915 and 1916, before the Motor Cultivator began production. Implements included grain shock loaders, various tillage implements, mowers, grain drills, corn planters, hay rakes, grain and corn harvesters, and grain headers, as well as methods for attaching other implements. E. A. Johnston identifies the Mogul general-purpose tractor as the true start of the Farmall tractor. Development of the tractor itself came to an end at the close of 1915.

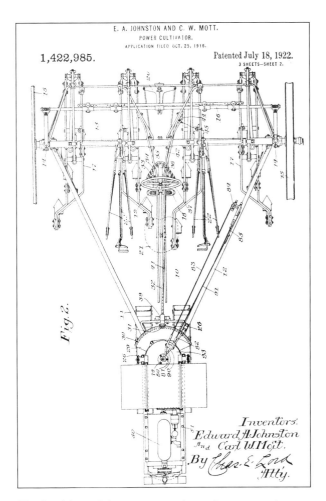

The first Motor Cultivator patent shows the two steering systems. The first system, used to make quick turns at the end of the row, used the steering wheel, which through a chain turned a shaft, Part 32. This in turn had a bevel gear at the other end which moved the whole motor and transmission assembly. The gear is part 31, while the rack for the gear is part 30. The other system, used in the row, used foot pedals, 18 and 19, to move two rods attached to a center piece, part 20. This, in turn, moved a tie rod which moved the front wheels at pivots 14 and 74. *U.S. Paten*

Motor Cultivator Development

Motor cultivators were literally a motorized cultivator. While a standard tractor could be used to pull many implements in a variety of situations, the Motor Cultivator could only be used for cultivation and belt work if equipped with a pulley. Motor cultivators were a brief but intensely popular fad of the late teens and early twenties.

International's first version of the Motor Cultivator was designed and built in 1915, the same year as the general-purpose Mogul, although the Mogul is mentioned as first in the histories. E. A. Johnston supervised the development. The first examples were completed at Tractor Works by late 1915. As first built, the Motor Cultivator did not have a drawbar, but one was quickly added. The Motor Cultivator is referred to as the 1916

The 1917 Motor Cultivator. The stability problem is starting to be dealt with by weights on the front wheels and by adding extensions to the rear wheels. The sign says "International Motor Cultivator—Two Rows at a Time." Other Motor Cultivators can be seen in the background. The photo was taken May 18, 1917. The 1917 Motor Cultivators were supposed to be getting into farmers' hands by this time, but were delayed by production difficulties. *State Historical Society of Wisconsin*

Motor Cultivator, as development really happened only in late 1915 and 1916. The first Motor Cultivator had drive wheels in the rear and cultivators mounted in the front, with the movement of the gangs connected with the steering apparatus of the tractor (horse cultivators had this shifting done by the feet of the rider). At the end of the row, the steering wheel was used to turn the rear wheel, which was locked forward when cultivating. Design was done by Johnston with the assistance of D. B. Baker, Carl Mott, Philo Danly, and others. These designers had in mind something quite different from the other motor-cultivators. They wanted the tractor to be able to do more farm work than just cultivating. Design of the general-purpose farm tractor seems to have been always the goal of the engineers, even while they were attempting to place a motor cultivator on the market.

While other companies were designing motor cultivators suitable for only this purpose, B. R. Benjamin, head of McCormick Works experimental department and an expert on adapting implements for use with tractors, was writing a memo clearly describing an all-purpose tractor. The memo, dated July 27, 1916, proposed a combined tractor and truck with three speeds carrying a harvester-thresher (IH's terminology for a combine), grain binder, a water tank and spraying device, and attachments for plowing, planting, disking, and other duties. While operating as a truck, the tractor had "2 to 4 ton capacity." The combined tractor-truck had four wide tire steel wheels, 9 inches or over, with a 15- to 25- horsepower engine with kerosene, electric, or other power.

According to Benjamin, all operations were to be performed largely by one person. Previous efforts to reduce the manpower necessary to operate tractor-drawn machinery had taken the form of ungainly remote controls. The ability of a farmer to drive horses either through reins or spoken commands, or through the training of the horses was taken into account in the design of horse-drawn farm implements. Most industry observers thought that a successful general-purpose tractor would have to use the existing horse-drawn implements because of the investment that each farmer had in them. This led to the common motor cultivator design, in which the farmer would be seated over the cultivators so the gangs could be foot operated.

Bert Benjamin points out another fact in his memo that would become very important in the Farmall program: the simplification and reduction of materials in implements that would be carried by the tractor instead of towed behind. Weight reduction by removing implement wheels and frames would become very important in years to come. Although not connected closely with the motor cultivator experiments going on at that time, Benjamin was clearly thinking heavily about the same idea.

Motor cultivator testing was conducted at Aurora, Illinois, during the summer of 1916. The 1916 Motor Cultivator could not pull implements quickly due to the steering system, which as mentioned above was foot operated until the end of the row and very slow to respond. The first motor cultivator prototypes overturned easily and wheel weights were quickly added to lower the center of gravity. This, in turn, slowed down

The Motor Cultivators put out for 1917 showed the need for yet further improvement. Apparently some difficulty had been encountered with weak front wheels, so heavier cast wheels were made. Although they were stronger and helped better balance the Motor Cultivator on side hills, they also made in-row steering much more difficult. The photo was taken December 28, 1917. *State Historical Society of Wisconsin*

This photograph, taken on December 28, 1917, shows a Motor Cultivator as it would be produced for the 1918 season. The wheels have been made heavier to avoid tipping. Unfortunately, the heavier wheels added weight and cost, but only reduced the tendency to tip, instead of eliminating it. *State Historical Society of Wisconsin*

the foot steering even more. Testing at Aurora also revealed that the cultivator, based on International's No. 5 horse cultivator, was too lightly constructed.

Production of the Motor Cultivator was debated in IH's New Work Committee, which oversaw product development. Several numbers for production were debated, with Johnston and several other IH executives favoring 500, one thinking 250, and two executives favoring the 100, which were eventually built. Johnston himself stated that there had been limited testing, and manufacturing large numbers was hazardous, but felt that IH had the combination of a radically new machine, "urgent" demand, the prestige of having the first successful machine on the market, and the ability to improve the financial standing of the farmer by eliminating the horse. Johnston's faction got at least most of their opinion, as an order was placed with IH's manufacturing department for 300 Motor Cultivators, with material ordered for 200 more. C. W. Gray wondered in his 1932 paper on IH's early history, *Notes on the Development of the Farmall Tractor*, about Johnston's view that the Motor Cultivator would be the first successful tractor in its class, as the Moline Universal and Avery

The Motor Cultivator on January 1, 1919. By now, a power lift (a worm gear unit located at the front of the tractor) and a power take-off unit (the power lift runs off this as well, with a shaft extending beyond the lift unit) have been fitted to the unit. This photo was taken at McCormick Works, where much of the early work on PTOs was done. *State Historical Society of Wisconsin*

IH next added a crawler unit to the rear of the tractor. IH was experimenting with standard crawler tractors at the time, but the purpose of this crawler is not increased traction, it is to lower the center of gravity of the tractor. The crawler unit placed the entire weight of the drive unit at a very low point. This tractor also had a power take-off shaft and a different power lift in the form of a winch, which can be seen mounted on the frame and is seen under the steering wheel in this photo, taken August 23, 1919. *State Historical Society of Wisconsin*

were already in the market and selling well, at least in the case of the Moline. Gray asked, "Was the wish father to the thought?" Gray knew that Johnston would be among the first to read his history and knew he had a temper. Yet Gray pulled no punches, perhaps the best indicator that Johnston made some very serious mistakes. Gray retained his job, even though Johnston was by then a vice-president of IH and was respected outside of the company.

Production of the Motor Cultivator was announced to the branch houses in November 1916. The writer of the memo, C. E. Allison, stated, "We are so confident of its success that we are willing to have built several hundred of them for the next season's trade." Allison stated that the tractor had been tested for "several weeks" and would retail for between $350 and $400. "Several weeks" was a very short time of testing (many of International's previous tractors had been tested for several months and some for two seasons) and $400 was far cheaper than any IH tractor then in production. Allison had been one of the two committee members favoring 100 tractors, but was now trying to sell the Motor Cultivators to the branches.

Something that Gray does not mention is the incredibly large size of the first order. IH usually first produced one or two prototypes of a tractor for initial testing. If successful, a preproduction order of several more tractors would be built and sent to various parts of the country for testing in varied conditions by real farmers. Problems would be fixed, often in the local blacksmith shop, or in later years by traveling IH machine shops that could build new parts in the cornfield. Rep-

A curious photo of a 1919 Motor Cultivator with a corn binder and bundle carrier. This photo also shows the PTO hooked up to drive the implement. The curious part is the radio—even IH engineers didn't remember what they were doing with this one a few years later. The photo was taken June 27, 1919. *State Historical Society of Wisconsin*

resentatives from IH's engineering staff or experts from the branches would check in frequently with the farmer to see how the tractor was doing. The amount of pre-production tractors varied over the years, ranging from 5 or 6 to 25 or so. The 1917 Motor Cultivator seems to have been an exception. Although the number of motor cultivators built in 1916 is a mystery, it couldn't have been more than two or three. Building 500 of a fairly unproven design was unusual.

Production of the Motor Cultivator must have run into serious problems. Only 48 had been built by June 21 and 58 by June 27. Since cultivating would have been in full swing in some areas, including the southern states, members of the sales department decided they had all they needed to fill the demand at the late

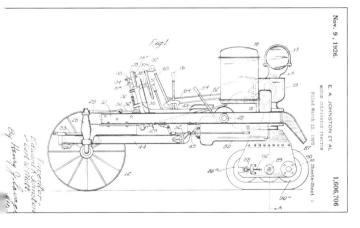

Of course, the crawler unit was probably a major reason for going to power steering. The early Farmall also had a subframe, part 44, which implements were attached to and could be lifted by power. *U.S. Patent*

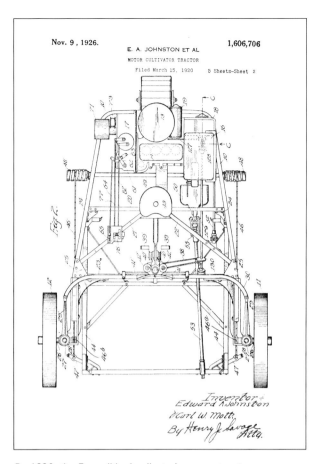

By 1920, the Farmall had collected power steering, a power lift, and a PTO. The PTO shaft is part 53, the winches for the power lift are parts 48, while the shaft from the steering wheel, part 129, leads back to the power steering unit, parts 18 and 19. *U.S. Patent*

date. One hundred Motor Cultivators were built before the cut-off. Only 31 farmers received Motor Cultivators in 1917. The unsold Motor Cultivators were used in demonstrations around the country. By September 19 all Motor Cultivators were recalled to Chicago to be rebuilt.

Reports from the farmers who had purchased Motor Cultivators were mixed. Although the Motor Cultivator did well in normal conditions on level ground, some farmers complained about the lack of power and speed. One farmer said that the Motor Cultivator traveled at 1 3/4 miles per hour, while his horses managed 3 miles per hour in the field. University of Illinois professors at Champaign cut the governor seal of their Motor Cultivator to reset the governor for more speed. Another farmer complained his Motor Cultivator could not pull a manure spreader that two horses could pull, even though the Motor Cultivator advertising claimed it could replace four horses. E. S. Beatty complained that the shovels were set too narrow to cross cultivate, but saw a different future for the tractor. He suggested removing the cultivator gangs and hanging a mower underneath, converting the Motor Cultivator into

Starting in 1920, gas power engineers started to radically change the Motor Cultivator. Now, power was being delivered at the two "front" wheels, removing the tendency of the castoring rear wheel to power the tractor onto its side. This is also the first tractor to be titled Farmall. Tractor wheels, instead of the previous cultivator wheels, were used on this tractor. This photo was taken March 19, 1920. *State Historical Society of Wisconsin*

a general-purpose tractor. Several IH executives saw this report. Newspapers also discovered the Motor Cultivator and were very complimentary, noting that the advanced machine was being operated by 15-year-old farm kids instead of factory experts. The Sales Department itself, while agreeing that the Motor Cultivator had done well in level ground, noted that the tractor was impossible to operate in hilly ground as the rear end kept sliding down and tore out the rows, and occasionally overturned.

1918

The New Work Committee approved 300 Motor Cultivators for 1918 production on August 22, 1917. Johnston noted that the LeRoi engine had done well but needed changes in the governor and fan. An example of the 1918 Motor Cultivator was supposed to be in the field the following week. The sales department thought that it could sell 500, but only if they were ready to ship by February 1918.

Advertising for the Motor Cultivator noted that the tractor had been designed for the cornfield and had a "strong drawbar for harrows, drills, seeders, land packers, land rollers" and implements that could be pulled by four horses. C. W. Gray noted that no mention was made of plows.

The main difference between the 1917 and 1918 Motor Cultivators was the 1918 heavy cast-iron wheels, aimed at reducing rollovers. The cast wheels increased cost and weight, which were already problems. Harvester's experimental departments worked on hitches and attachments for the Motor Cultivator to increase the number of jobs it could do for the farmer. Every horse the Motor Cultivator could eliminate improved the economics of tractor farming.

Unfortunately, the 1918 Motor Cultivator suffered the same problem of slow manufacturing as in 1917. Harvester officials probably did not spend a lot of money on tooling at this point, increasing the expense and time necessary to make the Motor Cultivator. One hundred and sixty were available for June 15, a far cry from the original goal of February. The production for the year was either 300 or 301. The sales department pressed for an early manufacturing order for the 1919 season in order to get Motor Cultivators on time.

Meanwhile, the manufacturing department did an audit to determine how much the 1919 models would cost to manufacture. On August 22, 1918, a report from the auditor said costs of $500 per unit could be "considered conservative." Given that sales prices were $642.50 for an 8-shovel Motor Cultivator, $650.00 for a 12-shovel, and $657.50 for a 16-shovel, International executives were concerned about the profit per unit. Out of the sales price would have to come transportation costs, sales costs, dealership commission (the dealers received a discount of 15 percent on the purchase of a Motor Cultivator, or $97.50), overhead, and the other usual costs. It was apparent that little profit, if not loss, would be left for International Harvester.

Another letter dated August 27 painted an even uglier picture of manufacturing costs. Other IH executives thought the manufacturing costs were too low and that IH would have to receive about $800 for every Motor Cultivator. They noted that the Avery Motor Cultivator sold for $540, and the market would only bear a price of $650 for the IH model. The order for the 1919 production of 300 was canceled on August 29. By March 1919, letters went out to all branches to liquidate the existing Motor Cultivator inventory.

The remainder of the Motor Cultivators sold slowly, despite the need for power on American and Canadian farms that arose during World War I. One hundred and fifty were left in stock at the start of 1919, while 62 were still left at the beginning of 1920. The last Motor Cultivator sold in the Mankato, Minnesota, area in mid-July of 1920.

The failure of the IH Motor Cultivator was due to several factors. The center of gravity in this tractor was very high, resulting in balance problems. Power, provided by a LeRoi engine, was not very great. The tractor was also very heavy for its horsepower and duties. Trying to fix any of these problems would have been nearly impossible given the layout of the tractor. Increasing horsepower, for instance, would have increased weight, and given the engine's placement, would have further raised the center of gravity and increased cost. Cost was the biggest factor in the demise of the motor cultivator. The motor cultivator simply could not provide all of the power needs for the average farmer, requiring yet another expensive tractor or horses. IH described the reason for the demise of the motor cultivator as, "It was found that it could not be produced at a cost which it was estimated the farmer would pay."

In *Notes on the Development of the Farmall Tractor*, C. W. Gray pointed out that most of the problems of the IH Motor Cultivator should have been discovered before the first prototype was even built. As part of his assignment to do the history (Gray normally worked in advertising) he was to write all that he learned. Any conclusions that he reached were to be reported. In return, he was not to be held responsible for the original draft (a second history was prepared later that had all references to internal politics and failures removed). In a few pages after the end of the Motor Cultivator section, he discussed vaguely the internal politics surrounding the Motor Cultivator. He seemed to place most of the blame on Ed Johnston's engineers. Gray said, "The Motor Cultivator tipped over. It was too slow. It was hard to handle. Its manufacturing costs were too high. And yet it took nearly six years for one faction in the engineering department to get it out of its system. And then it was because Mr. Legge had definitely placed his approval on another type of tractor." That tractor would be the lightweight Farmall.

In a 1934 letter, Johnston describes general-purpose tractor experiments as going on all during the years of the Motor Cultivator developments. While most historians believe that the Farmall was developed from the Motor Cultivator, the true history is that the Motor Cultivator was but one step on the road to the Farmall that began before the first Motor Cultivator was designed or built.

Other Possibilities

Other companies also developed motor cultivator or general-purpose tractors, some of which were offered to International for sale as complete projects. One such tractor was the Universal built by the Universal Tractor Company of Columbus, Ohio. This company built 50 tractors in 1913 and 100 a year later. In 1915 this company offered the Universal IH, which, after examining the tractor, turned the proposal down. Later in the year the Universal design was sold to Moline Plow Company, which further developed the tractor and sold it successfully for several years. It was the Moline factory (in Rock Island) that IH later bought for Farmall production in 1926.

Another tractor that the IH sales department was interested in selling was the Indiana, built by the Indiana Silo Company of Anderson, Indiana. The Gas Power Engineering Department sent out C. O. Aspenwall to investigate, but he reported back unfavorably.

Farmall Development

During production of the Motor Cultivator, and continuing after the failure, IH continuously evolved the design of the tractor and implements. One of the early

A side view of the 1920 Farmall, showing the tractor wheels, power lift, power take off, and chain drive. The expense of the "frills" was undoubtedly a major part of the high expense and weight of these tractors per horsepower. Again, a March 19, 1920, photo. *State Historical Society of Wisconsin*

ideas was the addition of a PTO. According to L. B. Sperry's reminisces, a motor cultivator was demonstrated at Savanna, Illinois, with a full line of implements mounted on the tractor and operated through the PTO. Little would come of the implement line and the Motor Cultivator itself. From 1916 until 1922, the McCormick Works experimental department, under the direction of Bert Benjamin, devoted half of its time to experimentation of the Motor Cultivator and Farmall implements. Deering, P&O (after its purchase), and other works also spent heavy amounts of time and money on the Farmall implement project, an enormous investment and gamble even for a huge firm such as IH. Of course, the Gas Power Engineering Department at Tractor Works was devoting heavy resources to the Farmall as well as the implement experimenters. Tractor-drawn implements had to be different from horse-drawn implements in order to accommodate higher horsepower and higher speeds. However, tractor-drawn implements, if designed correctly, also weighed and cost less than their horse-drawn equivalents. This cost and weight factor made tractor implements very attractive to IH's upper management, which was kept informed about weights of tractors and implements.

Benjamin's expertise in the tractor implement field was also put to use by IH's main competitor. Benjamin was sent to the Ford plant to assist in designing equipment for the Fordson. IH was supposed to cooperate with Fordson by providing implements to increase agricultural production during World War I. Some of this work may have actually been done at McCormick and Tractor Work. Fordsons have been seen in photographs at Tractor Works, unlike most competitors' tractors which were kept at International farms in various places, most notably Hinsdale. Further details of IH's and Benjamin's involvement with the Fordson are difficult to find.

A close-up of the drive units of the tractor in the previous photo. By now, the tractor is definitely a Farmall (at the time spelled "Farm-All"). This tractor is the first Farmall to have a differential, transmitting power to the final chain drive. It also had the differential brake, which basically used brakes on one side of the tractor to assist in turning. The frame looks as if it has been made considerably stronger. *State Historical Society of Wisconsin*

1919

The new Gas Power Engineering Department continued experimenting with the Motor Cultivator to see if it could produce a salable, workable product. One of the early changes was the name. The term "Farmall" appeared in the department's records on November 10, 1919. Johnston was apparently unhappy with the term "motor cultivator" and circulated a questionnaire around the department seeking new names. "Farmall" was entered by E. H. Kimbark.

One of the major goals was to increase the stability of the machine. The first method of trying to reduce rollovers was to add wheel weights to the main wheels. This had been an ongoing method first starting in 1916 and continuing into 1919. However, a new method was tried in 1919, changing the drive wheels to a driving crawler, the weight of which would increase stability.

Power lifts were developed in 1919, with patents finally being granted in 1923. One patent covered a reversible power lift while another covered PTOs, power lifts, and methods of attaching implements. Later in 1919 a method of operating the power lift through gearing from the PTO was developed. Equipment that used these new developments included grain drills, a binder attachment, mowers, a corn binder with bundle elevator, and a push binder (also referred to as a header-

The first photograph of a Farmall running with the engine and pivoting wheel in front. This is not the reversible tractor, but probably an experiment to test the idea of the reversible tractor. This tractor has open-type rear wheels, later popular in Texas. The photo was taken October 15, 1920. *State Historical Society of Wisconsin*

binder). Field work with these new implements was carried out with the tractor and implements at the Dunham farm in Wayne, Illinois.

Plows were fitted to a Motor Cultivator/Farmall for the first time. Side draft was reportedly a problem, and plows would be a dead subject for the foreseeable future.

1920

The year 1920 saw still more modifications to the basic Motor Cultivator, bringing it still closer to being a general-purpose tractor. One such change involved the rear end of the tractor. The official IH history states that this "may have been due partly to [IH president] Mr. Legge's insistence that whatever form the tractor took it must at least look like a tractor." This statement may be more than a simple preference for something conventional. Legge probably anticipated that the tractor could be a tough sell that conservative farmers might pass on if the tractor was too radical a departure from other competitive tractors. The motor was separated from the castoring steering unit, remaining stationary for the first time. The front wheels now provided the drive, with power transmission, through sprockets and chains. The differential included the automatic steering brakes which would show up on production Farmalls.

Nineteen-twenty also may have seen a change in Harvester's entire design philosophy concerning the Farmall. Previously, the implements were made to match the tractor, resulting in a workable tractor that might not necessarily be able to do much work due to the limitations of the implements attached to it. However, IH had started an extensive effort to gather information on actual farming methods, led again by Bert Benjamin. Gradually, IH began using this information to design implements. Tractors in turn were designed to fit the implements. IH's engineers felt that this was when true progress began in Farmall development.

Later in the year, the motor was turned, going from being a cross-motor to the more modern in-line direction. This particular tractor had a Waukesha engine and could be driven in either direction thanks to a reversible seat. However, this tractor was described as heavy, slow, clumsy, slow to steer while in the row, and poor at cultivating. All 1920 Farmalls had three speeds forward and reverse, again because they were reversible tractors. During this year, Benjamin sent another letter to Alexander Legge, putting forward a "Farmall Plan" that would eliminate horses on farms completely—a goal that finally came to pass decades later. This tractor was proposed to have two speeds forward and reverse.

International's executives and engineers were still recovering from the Motor Cultivator debacle. Members of the sales department were not interested in a new all-purpose tractor or they thought that IH was far from having a marketable model. Alexander Legge was

The next stage of Farmall development was the "reversible tractor," which ran in both directions. The engine was brought into line with the long axis, and the steering was brought under the tractor and simplified. The seat could be turned around so the operator could face into the direction of travel. This tractor was the first real step beyond the motor cultivator. The photo was taken November 9, 1920. *State Historical Society of Wisconsin*

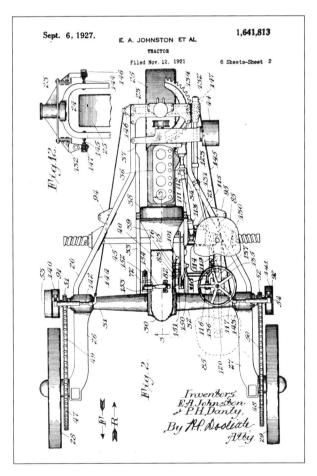

By 1921, the Farmall had acquired the independent turning brakes that would remain for years after. The brakes are parts 53 and 54, located just behind the forward wheels. This tractor still has the power lift and PTO, although it looks like the power steering has been relocated to directly under the steering wheel. *U.S. Patent*

Early Four-Wheel and Power Steering

One has to examine the patents IH received to realize how much technology IH was trying to put into Farmall development in 1920. One of the most startling developments that was experimented with was a steering system that consisted of two separate systems, one which was power steering and the other allowed four-wheel steering!!

The technology was developed during the struggle to make the Motor Cultivator viable. The Motor Cultivator's major problems were a high center of gravity and insufficient power. A larger engine solved the power problem, but worsened the high center of gravity. In order to lower the center of gravity, IH engineers looked into ways to add weight down low. One of the methods IH used to add weight at a low point was to use a crawler track rather than wheels.

Now remember, the engine pivoted with the drive mechanism during steering. With the crawler system, steering rapidly was a chore at the end of the row, where quick, sharp turns were called for. To solve the problem, IH devised a power steering system. However, steering in the row during cultivating was still done the traditional way born in the horse days—with the operator's feet. Foot pedals actuated the wheels at the front of the cultivator, where a small change was instantly translated into movement of the cultivator gangs themselves. IH then devised a steering system that used the foot pedals for minor corrections and a steering wheel for tight turns. The steering wheel is connected to a chain, which is in turn connected to a shaft, which in turn moves a clutch shaft. The clutch shaft had a screw machined into it that engaged a sort of clutch that fed power to turn the steering shaft. The operator had to continuously move the wheel to keep the clutch engaged to increase the angle of the turn. During the turn, the front wheels are prevented from turning by a pin, which dropped into the front steering yoke. However, when the tractor drive was turned more than 45 degrees, the pin in the front steering yokes was automatically lifted and the front wheels turned freely in order to shorten the turn—four-wheel steering.

watching developments but not pushing the issue. Johnston was tied up on 15-30 and 10-20 development. In fact, many people were looking at the 10-20 and 15-30 as the focus of IH's tractor program. There were other projects going on (the 12-25 and the steam program), but it was clear that the two tractors would receive most of the attention. In order to produce these tractors at a low price, an incredible investment would be undertaken to build assembly lines and high-volume tooling. This was in contrast to manufacturing in the past where the tractors were built by hand or by the progressive system, where tractors were assembled on hand-pushed carts.

There was one man who still had faith in the general-purpose tractor: Bert Benjamin. Ed Johnston kept P. J. Danly assigned to Farmall design, mainly as Benjamin's contact in the Gas Power Engineering Department. Benjamin was increasing his research into farm power needs, while Danly kept redesigning tractors to improve their ability to fit the need. Benjamin became an evangelist for Farmalls, writing memos and trying to convince the other decision makers of the possibilities. The proposed tractor could do more jobs (11) with one operator than the IH 8-16 (4) or the Moline Universal (9). The comparison with the 8-16 caused a rebuild of the Farmall. Power was made equal to the 8-16, with a transmission with two gears in each direction being added. Benjamin still supported the reversible tractor idea. Movies were made of the current Farmall for viewing by IH executives and shown for them December 13, 1920.

After the film, Johnston started the discussion, coming out on the side of the Farmall. Johnston stated that the tractor and attachments were not as complicated as many IH executives thought. He went further, saying that the Farmall and implements were less complicated than other tractors with regular implements, and that any discussion of the Farmall had to include both the tractor and implements as a unit when compared with other tractors.

H. B. Utley was on the other side of the issue. He saw seven problems with the Farmall idea.

1. There was a lot of machinery attached to the tractor which made it hard for the farmer to operate and maintain it.
2. The cost of buying the tractor and the implements to go with it was beyond the means of everyday farmers and suitable only for those not making their main living on the farm.
3. The tractor was too heavy.
4. Dual-purpose machines had never worked in the past.
5. Farmers would not scrap the machinery they already owned in favor of new Farmall implements, so introduction of the Farmall would be slow.
6. Due to complicated mechanisms, the Farmall would have lots of down time.

The next step in Farmall development was to move the cultivators to the "front end of the tractor." This 1921 tractor was also the first with gear drive, as can be seen by the rear position of the differential and the use of enclosed final gears. A rear-mounted corn planter is still connected to a power lift. This photo was taken May 23, 1921. *State Historical Society of Wisconsin*

Another view of the reversible tractor, showing an air cleaner, the frame, and a shifting arrangement for the cultivator gangs that was now added. The ridged rear wheels and lugs are also unusual. The location looks like one of the IH test farms, of which there were several at this time. *State Historical Society of Wisconsin*

7. Utley did not think farmers needed a fast-mowing tractor because the hay needed to cure anyway.

Utley saw some advantages to the current tractor, such as the PTO. He thought that the Farmall might be a good neighborhood machine, where several farmers joined to buy and use one tractor. He thought the sales department should have final say as to whether the work proceeded. A. E. McKinstrey doubted whether the Farmall could be marketed due to cost and novelty, but since the tractor had come so far, further experimentation should be continued.

Johnston spoke again. The engineering department was anxious to see the schedule for 1921. He stated a modest program would cost $150,000 and an active program $300,000, most of which would be tied up in actual construction of hand-built prototype tractors. After Johnston, the conversation among the executives centered around two questions:

Was the development going in the right direction?
Could a universal tractor be built?

J. F. Jones thought development was going in exactly the wrong direction, that a heavier tractor was needed, and that no one in his department was in favor of the Farmall. At this point in the discussion the Farmall was clearly in trouble.

Alexander Legge came to the rescue of the Farmall. He stated that there were many farm jobs to which IH tractors were not suited and that power machinery was still in a state of evolution. He thought it impossible to determine what the market would want at this point, and there was still the question of the all-purpose tractor to be answered. Legge clearly came out on the side of more work. After Legge's speech, Utley made a successful motion to build five experimental Farmalls, keeping the program alive.

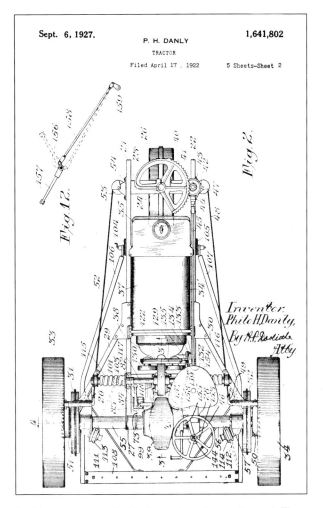

In this patent, several critical parts have been relocated. The brakes and final drives have been relocated, and the whole rear axle assembly has been tucked under the tractor instead of in front (or behind) it. The power lift is still there, but would be disappearing soon. *U.S. Patent*

Not much is known about this Farmall, except that this photo is probably from early 1921. IH's history of the Farmall has a curious gap during late 1921 and early 1922, although there were several different experimental tractors brought out. This seems to be a refinement of the "heavy" Farmall, as it still has the power lift. Still, it seems a smaller and more compact tractor than the tractors positively identified as the "heavy" 1922 Farmall, possibly indicating that the tractor engineers were downsizing their version as well. *State Historical Society of Wisconsin*

1921

Bert Benjamin's work at convincing other IH workers of Farmall possibilities was starting to pay off. More engineers became interested in the Farmall. Nineteen twenty-one would start to see real progress being made toward a real tractor. The major step came early. A new work report recommended altering the order for five experimental Farmalls to two lightweight Farmalls of a new design. McCormick Works engineers (probably C. A. Hagadone under the direction of Benjamin) had made a drawing of a small Farmall. This tractor represented a real change from the heavy Motor Cultivator-based Farmalls that the Gas Power Engineering Department had been working with. The implement side of IH had designed a tractor to compete with the tractor engineers. The cultivators were to be mounted at the front of the tractor. Having the cultivator in the front increased the "dodging" speed of the gangs, making the tractor better able to avoid tearing out the row it was cultivating. The engineers working on attachments were excited. Benjamin created some comparisons between the lightened Farmall and horses, and then the lightened Farmall and the Fordson. The comparisons revealed that a farm with the new Farmall could result in $3,500 of income for a 160-acre farm. The same farm with horses or a Fordson would pay the farmer only $3,000. Legge circulated this information to certain executives, referring to the new Farmall as "Benjamin's favorite tractor."

Benjamin had another proposal for Mr. Legge. He proposed manufacturing a Farmall attachment for the Fordson. By 1921 the Fordson craze was going full blast. Hundreds of thousands of the small tractor were in farmers' hands worldwide. Certainly, there was a market for a Farmall conversion. How the Farmall conversion would be added to the Fordson was a more difficult question. Apparently, Benjamin thought IH should be doing the conversions itself in IH factories. He reported that Ford was close to putting the same idea on the market and that IH should take the lead away from him. C. W. Gray compared the Fordson/Farmall as being very close to the F-12 in size but with different power. Benjamin worked further on the project as a possible attachment for the Fordson and patented the idea. The attachment was never built for sale.

By July 1921, the question of what to do about the Fordson was becoming more urgent. On July 21 a meet-

The "heavy Farmall" of 1922. This tractor is similar to the preproduction tractors that were sold to certain farmers (mostly IH executives and directors who owned "gentleman's" farms). By the time most of the 1922 Farmalls had reached the farmers, the design had already been dropped in favor of Benjamin's light Farmall, a fact that displeased some tractor engineers! This photo was taken April 17, 1922. *State Historical Society of Wisconsin*

ing was held regarding Farmall development and implement attachments for regular tractors. At this time, 240 Fordsons were being produced and sold a day. Henry Ford was developing implements for the tractor, and outside companies were selling equipment for the Fordson. To meet this competition, IH needed an active program to develop power-driven machinery. President Alexander Legge said money had been spent on power-driven machinery, but the sales department didn't like what had been produced. Legge said he personally didn't know where any of the 10 existing Farmalls were at the moment, and said that they were being followed closely by overprotective engineers when he could find one. Johnston answered Legge, saying that the engineering department had developed more power-driven implements than any other manufacturer, but that enthusiasm from other parts of the company had been low and no one was saying what kind of power-driven machines were wanted. Johnston thought the Farmall and attachments were better than the Fordson. After more discussion, some of it favorable and some not, it was decided to continue development, and to build 100 Farmalls for 1922. These Farmalls were to be built by hand and would be very expensive. IH executives were starting to warm to the Farmall, but the Fordson was starting to push them a little faster. The decision to build the 100 Farmalls was soon canceled and replaced by an order for one Farmall with a new cultivator attachment. A shifting cultivator gang had been devel-

oped that was connected with the steering gear. The shifting gear improved the response time of the operator dodging plants, and another part of the Farmall Triple Control System was in place.

The shifting cultivator gangs were not the only major change for 1921. Photographs dated May 1921 show the first gear drive for a Farmall. Gear drives were a major advance over chains, reducing wear from dirt and improving efficiency. IH was doing major work on gear-drive tractors in the 10-20 and 15-30, and the same technology was brought over for the Farmall. The ball bearing motors had been in Farmalls since 1920. In many ways, the Farmall benefited from IH's experience with the 15-30 and 10-20.

International company executives were becoming even more interested in the Farmall. McKinstry sent C. O. Aspenwall to Iowa to check on some Farmall experimental tractors. Aspenwall was told that his

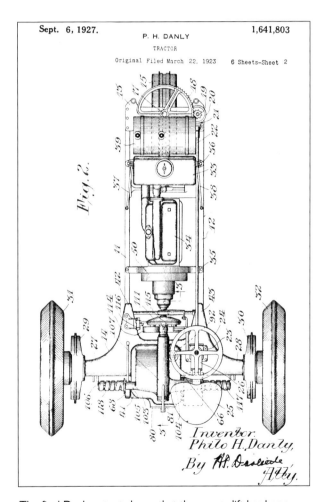

The final Danly patent shows that the power lift has been moved to the rear of the tractor, while the front body has been considerably narrowed. By the time this patent was applied for, Benjamin's tractor had swept aside the Danly "heavy Farmall," but presumably there were parts of the design IH wanted to secure for itself or keep out of the hands of others.

This is the small wooden model used by the McCormick Works experimental department to help design the layouts for Farmall implements. Here, they are examining how to position a check-row corn planter so that it runs in the wheel tracks of the Farmall. This photograph was dated March 20, 1922. *State Historical Society of Wisconsin*

The tractor that did it all—the "lightweight" Farmall. This is the tractor that convinced Legge that the Farmall should be built, convinced the other negative executives, and set the pattern for what ultimately entered production. This Farmall was the product of Bert Benjamin's push for a smaller, lighter tractor. The power lift has been eliminated; the tractor now is not a reversible tractor—the seat is fixed and there are three speeds forward, one reverse. Although a sheet metal hood is added, the tractor is still much lighter than its direct predecessor. This photo was taken July 19, 1922, less than a week before the conference that decided to go with this tractor and abandon the heavy Farmall.

report would be very important to the future of the program. Aspenwall, who had done similar reports before with the Indiana tractor as well as others, reported back that the Farmall idea was "sound." What was needed was a tractor with a correct design.

The IH New Work Committee took another try at deciding how many Farmalls to build for 1922. The single Farmall built with the front-mounted cultivator had done very well at Hinsdale. Many numbers were discussed, but 20 were finally authorized, with 20 cultivator attachments, 20 Farmall mower attachments, 20 Farmall corn planter attachments, 10 Farmall binders, and continued development on several other attachments.

According to the IH historian C. W. Gray, Benjamin had started to become discouraged with the slow pace of Farmall development during 1921. Other work was producing results. The Farmall program had taken enormous amounts of money and produced nothing. Benjamin thought he was isolated, with only Harold McCormick as a supporter. Alexander Legge was following the program closely and noted Benjamin's problems. Legge sent Aspenwall over to Benjamin, saying, "Charlie, for Heaven's sake, go to Benjamin and see if you can give him some help and encouragement. He's got a real job on his hands and we want to see him succeed. Do anything you can to help him work it out."

Legge missed few things in the IH organization and was loved by many employees. It is certain that without him, the Farmall project would have died.

1922

In 1922, the Farmall was transformed into a real tractor. Still, opposition within the IH organization would have to be overcome. The Fordson wars were putting heavy pressure on IH's executives to come up with a cheap tractor. The 10-20 wasn't in production and the 15-30, while being produced in small numbers, was a much larger tractor. Meanwhile, work was continuing on the 20 Farmalls authorized in late 1921. Seventeen of the reversible tractors were produced and put into the hands of farmers. Many of these tractors went onto the farms of IH executives. Most of the executives were "gentlemen farmers" who employed managers and staffs to run their farms. However, Alex Legge was notorious for doing as much farm work himself as possible. Guests arriving at his farm were often asked to pitch in with the chores, including shoveling manure. If they were wearing fancy clothes, they were given boots and coveralls. Legge applied this treatment to all guests, including U.S. presidents.

At some point in 1922, Benjamin, despite the encouragement he was getting from Alex Legge, got tired of the internal politics involving the Farmall. Gray,

The 1923 Farmall, looking definitely like a Farmall then. The tall air cleaner (designed to take in air from out of the "dust zone" around the tractor), the rear transmission case, and rear wheels look like the 1924 version, but the front bolster and steering wheel would be changed, as would the upright exhaust and manifold. Several of these tractors were sold to real farmers for real testing. This photo was taken February 26, 1923. *State Historical Society of Wisconsin*

Bert Benjamin

Bert Benjamin represented the next generation of IH engineering from the "Experts" and "Experimenters." Benjamin was one of the first college-trained engineers hired by McCormick Harvesting Company. Benjamin was born and raised on a farm near Newton, Iowa, and graduated with a degree in mechanical engineering from Iowa State College. After graduation, he got a job at the McCormick Harvesting Company and stayed until retirement in 1940. He became chief inspector of the Farmall Works in 1901. He became the superintendent of the McCormick Works experimental department in 1910 and assistant chief engineer of Farmall Development in 1922. In 1937, he was assigned to research in agricultural practices, a subject that was always near to his heart and that had served him well in the battles over Farmall development. While at McCormick and at Harvester he received over 140 patents.

Benjamin was always known as a friendly man, but one who was unafraid to fight for what he knew was right. Despite becoming an executive with an office in the Harvester Building in downtown Chicago, he was often found in a small building at the McCormick Works, still engaged in hands-on development.

After his retirement, Benjamin remained very active, swinging Indian clubs daily for exercise. He remained interested in engineering until 1969, when he broke his hip. A few days later, he died in his home in Oak Park, Illinois.

---

the man IH assigned the task of writing the Farmall history years later, tried to dig into what happened, but was forbidden to talk with Benjamin or Ed Johnston about the curious events surrounding the Farmall in 1922. Reading between the lines in Gray's history (if he printed the truth he probably would have lost his job even with the protection he got as part of his assignment), there was probably a blowup between Ed Johnston and Bert Benjamin. Johnston was in charge of the Gas Power Engineering Department, which was normally in charge of tractor development. Benjamin was in charge of the McCormick Works experimental department, which designed implements in the McCormick line, other implements built at the McCormick Works, and agricultural implements in general. However, Benjamin was really pushing the Farmall and started to use his staff on tractor design. There is a line drawing from McCormick Works that shows the first glimpse of what would become the Farmall. It was a nonreversible, tricycle tractor. The caption on the drawing says, "Combined Features-Plan April 1921 and December 1921. International Harvester Comp. McCormick Works." Although it is not clear when the drawing was made, it's clear where it came from: Benjamin's McCormick Works experimental department.

Johnston apparently was not happy with this invasion of his territory and started to oppose the Farmall program (or at least withdrew his support) in retaliation. The office politics got to the point that according to rumor, Benjamin threatened to quit International Harvester unless he was given free rein to pursue the Farmall without interference. Johnston, while respected and loved by his employees for his brilliance and fairness, had a legendary temper. A few stories tell of Johnston chewing out employees for extended periods of time, while others wondered how his assistants survived. In Johnston's defense, he worked extremely long hours and is known to have suffered extreme pain from rheumatism. Nonetheless, he was not a man to pick a fight with. Benjamin did, and won, gaining free rein (within reason) and control of Farmall development.

A special conference on the tractor program was held February 10, 1922. The possibility of bringing out a cheap tractor had been discussed in Springfield, Ohio, the previous week. Benjamin suggested a cheap tractor could be made from the Farmall by reducing the power, reducing rpms from 1,250 down to 1,000, and building it to handle one plow instead of two. He thought that this would bring weight down to 2,400 pounds and suggested gray iron gears and cast wheels to reduce manufacturing costs. Other executives thought that concentrating on 10-20s or other small conventional tractors would be better. Still, they thought that they should allow Benjamin to build one or two small Farmalls to test the idea.

The pressure of the Fordson and the slow pace of the Farmall were even starting to get to Legge. Legge wrote to Utley that the Farmalls in the field were poorly put together and suggested that perhaps a stronger engineering staff was needed to prevent fur-

This is a photo taken during the 1925 meeting between IH executives and Texas dealers. At this meeting, the Texas dealers threatened to form their own company to build the Farmall if IH did not continue with the project. Bert Benjamin is identified as the fellow on the far right. *State Historical Society of Wisconsin*

ther problems. Testing of the Farmalls also bothered Legge. The Farmalls were being followed by IH employees closely, preventing an honest opinion from the farmers. He considered the Farmall too complicated and a long way from being ready to put into production. Without changes, 1922 would end with no improvement and put the future of the Farmall into doubt. In his history, Gray, the historian, wondered if Legge was just trying to get the engineering department motivated with his letter. Gray notes that this was the only evidence he ever found where Legge was pessimistic about the Farmall. However, the day after Legge wrote his letter, he would be in a completely different mood. An answer had been found after all those years.

A new Farmall at Hinsdale was demonstrated for Legge. Reporting back to the executive council, Legge stated that Benjamin's small machine was a big improvement. He felt that this tractor could be followed until a successful tractor was marketed. The small tractor had to be able to cultivate and handle row crops, which Utley said it would do. Utley stated that the tractor could also pull 12-inch bottoms. Utley left directly from the meeting for Hinsdale to look at the new Farmall, but before he left it was decided that all development on the heavier Farmalls would be stopped.

The new tractor looked like a tractor. The idea of a reversing tractor had been abandoned. Almost every area of the tractor was changed. By July 29, IH decided to stop all expenses whatsoever with the 20 heavy Farmalls in the field. If the farmers wanted to continue using them, however, they could.

The sudden stop of the heavy Farmall program did not make the tractor engineers happy. L. B. Sperry thought the decision was hasty. He wrote Johnston, suggesting that statements be gathered from the farmers currently using the heavy Farmalls. Sperry thought that the sales department had killed the heavy Farmall before the tractor had even made it off the drawing board. Other executives, meanwhile, thought the Motor Cultivator should be brought back in a one-row version. These ideas quickly died when reaching the upper IH executives, who were sold on the light Farmall. The light Farmall even won over the hardest critic, J. F. Jones. Although they did not consider the Farmall ready yet for production, and a few executives were still negative, most of the IH executives knew they were close.

1923

An executive council meeting on February 23, 1923, saw Legge telling the council that they were closer than ever and that the question was how many to build for 1923. Building a large number of Farmalls would be difficult since Tractor Works was now building 10-20s as fast as it could. The engineering department would have to build the 1923 Farmalls by hand, instead of the manufacturing department. Mr. Everson was not satisfied with the Farmall, and thought the Motor Cultivator was still the best machine for the jobs the Farmall was planned for. He also said that there was a strong prejudice against three-wheel tractors (which ironically IH had helped establish when the competition was selling three-wheel tractors!). Johnston pointed out that Farmall mowers could be built for $16 versus $34 for a regular mower. In any event, the new Farmall could pull anything the farmer already owned. Utley was afraid that the Farmall could hurt 10-20 sales, but would face that situation when it came up. Even further conversation took place, but a unanimous decision was reached to build a maximum of 25 1923 Farmalls with no radical changes to be made in the tractors. Johnston told the committee that the tractors would start coming out in the middle of May, with the last one available in September. Production of 25 Farmall cultivators and 20 mowers started June 15 at McCormick Works.

The Farmall was thought by some as a cheap tractor to compete with the Fordson. Production costs and projections kept coming back that the tractor was not going to be much cheaper than the 10-20. Legge wrote about that subject to Johnston, calling Johnston's attention to that fact. Legge also was not happy with an attempt by the engineering department to build some of the Farmalls with a different type of rear drive. The time had come to stop the experimentation according to Legge. Six years of experiments had come and gone, and now the organization needed to become familiar with the tractor to sell it. Changing it now would interfere. Legge's letter killing different types of rear drives for the Farmall was a little unfortunate because production Farmalls in the future would have serious problems with transmission and rear drive breakage.

By August 23, 1923, 26 tractors had been built. Thirteen of the 26 were being experimented with, while another 13 had gone out to the branch houses for demonstration and possible sale. With the exception of one tractor, all were doing well. IH engineers were following the tractors. Benjamin traveled the South examining farming conditions and experimental machines, especially the tractors engaged in cotton growing. The Farmall and tools for cotton, including dusters and middle breakers, were showing that they could save cotton growers money. One tractor was saving $10 an acre over using horses at a time when $10 was quite a bit of money. Another farm manager said that the new Farmall beat the Titan "all hollow" while filling a silo. More than one report said the Farmall had "plenty of power." A report from Madison, Wisconsin, said that "a 10-year-old child could easily steer it."

There were some complaints. Certain parts would have to be strengthened. The crank for starting the tractor was connected to the crankshaft by gears, which were troublesome. Cultivator gangs bumped into crankcases. However, there was nothing major to be found wrong with the tractor. In fact, at least one of the farm managers presented Benjamin with a written order for three more Farmalls (unsuccessfully).

On October 9, 1923, the IH executive council raised the question as to how many Farmalls to build for 1924. Mr. McKinstry was of the opinion to build 100. While the users were happy with the tractor, he felt the introduction of the tractor in larger quantities would hurt the sales of 10-20s. In addition, prejudice caused by the unusual appearance of the Farmall had to be over-

Although IH had experimented with power lifts in the early Farmall, another company put the power lift into production. By 1929, the demand for a power lift was starting to grow, so IH started experimenting. This is a worm drive power lift running off the PTO. The worm drive was serial Q-885, photographed December 14, 1929. *State Historical Society of Wisconsin*

come, and the similarity of costs between the Farmall and 10-20 would hurt the 10-20 as well. Besides, McKinstry thought the tractor could use another year of watching. Legge and Johnston were in favor of going further than this, as long as the expense could be held down. The 1924 Farmalls would be built by the manufacturing department rather than the experimental department. It was decided at the meeting to build 200 Farmalls, if the accounting department found that they would not be too expensive.

1924

While 200 Farmalls were authorized, 205 were built and shipped to the branch houses for sale. IH did not do extensive advertising on these tractors. There were only a small number to be sold, and the company was anxious not to interfere with 10-20 sales so that the massive expense of tooling up for producing the tractors could be made back. The 1924 Farmalls were still built by hand. The price of $825.00 for the Farmall and $88.50 for the cultivator attachment came nowhere near the cost of producing the tractors, the loss being chalked up to the cost of introducing the tractors.

The 1924 Farmall, while looking similar to the 1923 version, had numerous changes, including a one-piece, strengthened bolster (the part at the front of the tractor holding the steering gear), increased diameter in the rear axle, increased strength in the drawbar fastenings, a different front wheel, steel steering gears, an oiler for the rear axle roller bearing, strengthened bull gear

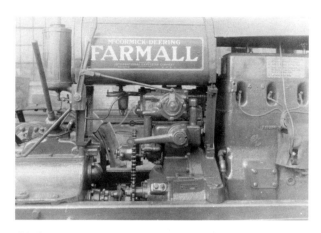

IH also started experimentation with a hydraulic lift on Farmalls. This was a compact unit running off the shaft between the engine and transmission. The control lever can be seen next to the decal. The hydraulic unit was serial Q-846 and was photographed October 4, 1929. *State Historical Society of Wisconsin*

housing, and a different main frame. The gas tank was increased in size, while the tractor shipping weight had increased 21 pounds.

The first order for the Farmall was shipped February 9, 1924, to Taft, Texas. The name of the farmer who received QC-501 does not appear in the Farmall histories done by IH, but it's obvious that he played a very important role in developing the Farmall. This same unknown farmer requested IH make the first set of skeleton wheels for the Farmall to deal with his soil. More important, he also pioneered the use of the four-row cultivator. The first year he had the tractor, he used the two-row cultivator and pulled two two-row riding cultivators behind. The next year, IH built and tested the four-row cultivator, with some of the work being done in the Cage Implement and Hardware Company blacksmith shop in Taft. This dealership would sell more than 90 Farmalls in 1925, over 10 percent of production.

Roy Murphy of Iowa was the first to actually receive a Farmall—he received QC-503 (it took time to get a tractor to Texas, as opposed to Iowa, in the 1920s). Murphy's Farmall became famous as "the first one." In later years, he often brought the tractor to Farmall Works for promotions and anniversaries since he lived nearby in Iowa. Eventually IH bought the tractor back from Murphy.

Reports started coming in from farmers about the new tractor. One farmer reported discarding his Fordson after starting with the Farmall. At least one farmer wanted three or four more Farmalls. However, a trip by H. F. Perkins to visit nine different Farmalls revealed that some redesign was necessary, although the tractor operators were very enthusiastic. More Farmall implements were needed. Most of the Farmalls produced went to the Central and Northwestern regions of the United States, although some made it to nearly every part of the country. On November 18, the price of the Farmall was increased to $950 to test the market and to take less of a loss on manufacturing.

1925

Since the 1924 Farmalls had been so successful, more were planned for 1925 to test the market. Numbers would be increased, but they were still limited by a lack of factory space. Although 250 were authorized for 1925, this number would be passed quickly due to demand. The first 1925 Farmall, QC-701, was finished either October 26 or November 6, 1924, with the last being built February 21, 1925.

Experience with the 1924 Farmalls had proved that further development was necessary. Changes were many, including

- Simplified steering crank mechanism
- Increased thickness rear axle flange
- Simpler steering gear
- Another redesign of the front bolster with fewer parts
- Stronger and larger rear wheel parts
- Gray iron front wheel hubs with "T" head spokes
- Strengthened drive gear hub
- Pomona air cleaner
- Ability to attach rear fenders to the rear wheel carrier plate
- Muffler instead of exhaust pipe
- Simple gearshift
- Simpler transmission
- Stuffing box for the belt pulley
- Repositioned spark and throttle levers (to steering shaft bracket instead of fuel tank bracket)
- Simple mounting of clutch pedal
- Stronger transmission case

Although only 250 were authorized, 838 were built. By November 1, 250 were shipped to Texas alone, a big increase from the 13 tractors shipped there in 1924. One farmer from Houston, George Newgent, had replaced 90 mules with 5 Farmalls. On one occasion he ran one of them six days, only shutting down to refuel. Eventually, the order went in to IH's manufacturing department for 1,000 Farmalls for 1925, but not all could be produced.

At this point, International executives realized that the Farmall was not interfering with 10-20 sales. The early Farmalls were being taken largely by cotton-growing states, while the 10-20 had sold very little in these areas. The Farmall was a godsend for cotton farmers. According to Benjamin, cotton in 1925 cost the average farmer $110.00 a bale to grow with horses or regular tractors. With the Farmall, that price was reduced to $83.33. The cost to grow that same bale of cotton overseas with "cheap labor" was $95.00 a bale. The Farmall was the difference between making a profit or a loss. Benjamin thought that it would take 300,000 Farmalls in eight years to replace 2,500,000 mules in use at that time.

The Texans, meanwhile, were doing a lot of work in designing and building Farmall implements. Four-row implements were being called for, and IH dealership machine shops in Texas were experimenting and building them. The dealers had also discovered a very effective way of selling Farmalls, starting in 1924. If a sales prospect didn't think the Farmall could handle the job, looked funny, or generally was resistant, the dealer would loan him a Farmall for a few days to try out. Usually that was enough to make the sale. If the farmer was still balking, the dealer sent out an employee to pick up the tractor to take it back to the dealership. That ended the argument in favor of the dealer. According to C. W. Gray, not one farmer successfully resisted the above method in 1925.

Texas pretty much settled the question of whether to build the Farmall. IH was still hesitant to build more in 1926 and sent an executive south to discuss the issue. In a meeting held in a Texas field, while watching the Farmall operate, the Texas dealers gave IH an ultimatum: build the Farmall, or the dealers would combine and take over the Farmall project and build it for themselves. The executive went back to Chicago and reported the prospects were good for future sales.

With both the Farmall and 10-20 doing well, the major problem was building the tractors. The 10-20 was well established at Tractor Works, having devoted assembly lines that were building thousands of tractors nearly full time and squeezing in Farmalls besides. Milwaukee was building 15-30s and engines for other IH products. It was proposed that Farmall production be started at Deering Works, but the idea probably didn't get far. Deering was from a different era of small horse-drawn implements and would be shut down by IH in 1930. The problems in production meant that no advertising was done as the market would take more than IH could produce at the time. Sales were still being withheld from areas with strong 10-20 sales, so even more demand was possible. Demonstrations in International's Omaha sales territory had IH dealers begging for Farmalls.

Farmall production went, of course, to a factory that IH had purchased cheaply from a company having severe financial problems. The factory was the old Moline Universal tractor factory in Moline, Illinois. Originally called the Tri-Cities Works, the factory would soon be renamed the Farmall Works.

While some development would be needed (especially the transmission, which had a history of breakage) the Farmall had finally come to be a major product for IH. Still, IH would return to many of the features that were experimented with and discarded during the experimental process, such as power lifts.

A 1924 Farmall with golf course wheels seen at an Aurora, Illinois, country club. Not only was the Farmall pulling rollers, its own wheels were acting as rollers. This experiment eventually led to the Fairway Farmalls. This tractor was photographed September 12, 1924. *State Historical Society of Wisconsin*

The Fairway Farmall

It may seem strange to sell a cultivating tractor to a golf course, but International Harvester did. The Fairway Farmall had flat broad wheels, but was otherwise a standard Farmall. The Fairway Farmall was not used only on golf courses but also shows up frequently on airfields in IH's advertising.

The reason for using the Fairway, as opposed to a 10-20 with the same wheels, is in the layout of the wheels. Rolling the grass is a common duty on golf courses. Rolling is simply pulling a roller over the turf to keep it firm. The tractor itself can roll the grounds while performing another duty, such as mowing. On a standard 10-20, the front and rear wheels lined up with each other, rolling some of the course twice while leaving an unrolled gap in the middle underneath the tractor. With the Fairway Farmall, the tricycle front, fitted with broad flat turf wheels, rolled the space between the rear wheels, leaving very little gap when covering the course. Airports in the twenties and thirties had grass runways that needed to be rolled, which was done by the Fairway's wheels as it mowed the grass. Fitting a drawbar to the airport tractor was a natural as the size of the aircraft increased during those decades.

Originally, the golf course tractors were just ordinary Farmalls fitted with the special wheel equipment, and at least one 1924 Farmall was fitted with the special equipment. The original golf course tractors had a "golf course wheel attachment" that was authorized for standard tread Farmalls February 5, 1926, although the request for authorization had been made December 21, 1925. IH also made golf course wheels for narrow treads. IH probably instituted the term "Fairway" as a marketing tool in later years.

This is the first orchard attachment for the Farmall. During the 1920s and 1930s, many experimentals were constructed for orchard use. International offered the orchard fender attachment for sale in the early 1930s, but few were sold. The option added 189 pounds to the tractor's weight. This photo was taken August 6, 1931. *State Historical Society of Wisconsin*

Narrow-tread Farmalls

Narrow-tread Farmalls were first authorized in 1927. However, the need for the narrow tread probably had been with the Farmall program since the beginning. There were dozens of different row crop spacing in the South, and IH's executives had been discussing the question. However, Alexander Legge, in a trip through the South in 1923, saw only a few of these row spacing and generalized that his engineers had been blowing the problem out of proportion. Back in Chicago during a meeting about Farmall development, Legge told the engineers that the row width was not a problem in the South and to proceed accordingly. Ed Johnston vigorously protested, but Legge's mind was made up (Johnston probably wasn't too popular at this point either). However, the truth was unavoidable, and development of a narrow-tread Farmall was made necessary, although the product decision states that the tractors were "for the Argentine." Production of narrow-tread Farmalls was authorized March 10, 1927, with Tri-City Works being ready to supply the tractor on order May 5.

Industrial Farmalls

There is an interesting reference to industrial Farmalls in one industrial tractor sales document dating from the early 1930s. The document refers to about 300 Farmalls having been sold for industrial purposes to original equipment manufacturers (OEMs). OEMs traditionally bought tractors (or major parts of tractors) and converted them into a piece of equipment that was sold under the OEM's name. Tractors going to OEMs were sold to the manufacturer directly by IH and were not necessarily advertised by IH, although many pieces of equipment were advertised and sold through IH Industrial Power dealerships. The author was unable to find any information as to who used these industrial Farmalls in their equipment, but the reference stated that the Farmalls were used for sweepers and other equipment.

Orchard Farmalls

Orchard tractors are ordinary small, very low-profile, streamlined tractors designed to go under and next to fruit trees in order to cultivate the ground surrounding the trees. Farmalls, however, are very tall tractors that do not go under low-lying branches very well. Why, then, convert a Farmall into an orchard tractor? Using a row crop as a cover crop between the trees is a definite possibility. During the first few decades of the 1900s, there was much experimentation as to what crops would be best grown between the trees in an orchard. Some farmers tried row crops between the trees, including corn. A tractor such as the orchard Farmall would have had to have been used to cultivate the rows, especially if the farmer got a little close to the trees.

Gas power engineering photos reveal that the orchard fender attachment was being developed starting around 1928. The attachment was authorized on November 16, 1931, but was discontinued due to lack of sales in March 1935. The attachment was designed for the Farmall narrow-tread tractor, but was not advertised well (if at all) by IH. The attachment was probably sold through select branches in regions with large numbers of orchards.

Chapter 6

The Improved Power Program

A new wave of development came from the Gas Power Engineering Department in 1930. References enter the photo albums about the "Improved Power Program." The photo albums make very clear the path of development and the results, the Farmall F-20 and F-30, and the W-30 and W-40.

There are several clues as to why IH decided to remodel its product line at this time. The first is time itself. Most of the engineering in the 15-30 and 10-20 series was now 10 years old. Since then the rate of improvement in technology dealing with tractors and gas engines was tremendous. The second clue is competition. The tractor business had staged a partial comeback from the terrible IH-Ford war and new, more advanced tractors were coming out throughout the industry. Oliver had been testing a Farmall competitor in 1928, but it was J. I. Case that came out with the Case C and L in 1929 and the CC in 1930.

The Case CC in particular was a scare for IH. IH records clearly indicate that the CC was the first real competitor for the Farmall. (Although the John Deere GP row-crop tractor had appeared in some forms earlier, IH considered the date a tractor was tested at Nebraska as the official entry date. IH really didn't consider the GP as much competition anyway.) The CC's horsepower rating was much larger than the Farmall while Case's unit construction ensured a much lower weight than IH could build with the frame construction of the Farmall. Case's L and C were also powerful, well-engineered tractors, although available IH records do not specifically mention them as a problem. The combination of the above probably explains the need for new IH tractors, especially a larger Farmall to compete with Case. Other competitors were also close to bringing out row-crop tractors, including Oliver (a tractor with some design similarities to Case), Allis-Chalmers, and a weak effort by John Deere.

For the Farmall series, two series of development were followed. First, the Farmall Regular was modified to increase the power from the basic engine. A vertical manifold was, of course, the most obvious feature, but the transmission also received attention, as well as air cleaners, carburetors, magnetos, engine and radiator screening, and other components. This redesigned Farmall tractor became the Farmall F-20

The F-20 started out as an improved power Farmall Regular. The Farmall Regular had been around since 1924. By 1931, other companies were catching up, and IH needed to keep pace. The improvements to the Farmall included a different manifold system (exhaust to the top), increased horsepower, and other small changes. Tractor Q-1549 weighed 4,248 pounds. This photo was taken March 17, 1931. *State Historical Society of Wisconsin*

An experimental wide front axle on a Farmall Regular. One of the advantages to the F-20 development program was that new parts and ideas could readily be tested on the Regulars. The new front axle weighed 444 pounds. The photo was taken April 3, 1931. *State Historical Society of Wisconsin*

The first experimental F-30 was titled the "Redesigned Farmall Tractor Using 4 1/4 x 5 Engine." It was basically a scaled-up Farmall, using a larger engine. This engine has a very interesting manifold with what looks like a valve to divert exhaust heat to the intake manifold to assist in vaporizing kerosene. Tractor Q-1029 was photographed May 3, 1930. *State Historical Society of Wisconsin*

and entered production in 1932. However, there is evidence that there were at least some Farmall Regulars (as they were soon called) built up to 1936. The reasons for the continuance is unknown, but is probably due to certain attachments that were designed to fit only the Regular.

The F-30 itself was redesigned, and the new tractor is shown in this photo. A new, simpler manifold with vertical exhaust cleans up the look of the tractor. New front and rear wheels, a new toolbox, and changes to the steering and air filter are also apparent. Tractor Q-1382 weighed 7,210 pounds and was photographed December 29, 1930. *State Historical Society of Wisconsin*

This is one of the first F-30's narrow treads with a wide front-end attachment. By this time, IH was familiar with the market for narrow-tread tractors. However, they did not take the more obvious path of going with the J. I. Case CC type of construction without side housings and final drives for this tractor, as they did with the F-12. Tractor Q-1739 weighed 4,880 pounds. The tractor was photographed August 11, 1931. *State Historical Society of Wisconsin*

Farmall F-30

The second path was a Farmall tractor built around a larger engine. This tractor, the direct competition for the Case CC, was the Farmall F-30. Unlike the Case tractor, the F-30 was designed from the ground up as a row-crop tractor, instead of being a variant of a conventional tractor (the CC was based on the Case C). This wasn't quite as much of an advantage as one would think. The all-unit construction of the Case resulted in a much lighter tractor than the IH competition, while still remaining powerful.

The scaled-up Farmall also saw experiments in transmissions, hydraulic and power lifts, and wide front ends (which were also experimented with later in the Improved Regular program). Both the F-20 and F-30 tractors saw experiments with various lighting systems, both with and without batteries.

For the tractors, the development proceeded first along the lines of improving the 10-20 and 15-30. Vertical manifolds and other assorted improvements were tried. The modifications to the 10-20 chassis included a new front axle (that was in some cases fitted to 10-20s that were still being produced). The new tractor was named the W-30 and was authorized for production December 8, 1931. The industrial version, the I-30, had certain differences, such as a four-speed transmission instead of the normal three-speed. The I-30 was more adaptable for mounting equipment than the previous small industrial tractor, the Model 20. It also had a more rugged drawbar. When fitted with pneumatic tires, the W-30 could be ordered with the four-speed in certain territories. Production of the 30 series of tractors began May 24, 1932, on tractor IB 512, which was an I-30.

The improvements for the 15-30 followed the same path as the 10-20 for quite some time, but then radically changed. IH's diesel engine experimentation was by then producing engines capable of being run in farm tractors, and a few diesel-powered 15-30s were tested, mainly in Arizona. But the increased power gasoline version of the 15-30 engine was replaced by a six-cylinder truck engine (the Gas Power Engineering Department was responsible for truck engines as well). The efforts required to adopt the diesel engine and the six-cylinder truck engine caused the revamped 15-30 to be delayed until 1934, when the new W-40 series was announced. The tractors comprising the W-40 series were the WD-40 (diesel), the WA-40 (gasoline), and the WK-40 (kerosene). The kerosene and gasoline versions had slightly different engines.

Diesel Development

Diesel development began in 1916 or 1917. When diesel development began, most diesel engines were huge monstrosities weighing several hundred pounds per horsepower and suitable for large stationary plants or ships. However, the diesel engine had two advantages that could not be ignored: the use of low-cost fuel and extreme efficiency in using that fuel. On the downside

The W-30 started out as the increased power version of the 10-20. In this photo, the revised manifold and other engine changes are plainly obvious. This 10-20 gear drive (increased power) is tractor Q-1435, weight 4,310 pounds, photo dated December 29, 1930. *State Historical Society of Wisconsin*

were weight, complexity (especially in injection), mechanical accuracy required, and the resulting expense of the above factors. The equation of the diesel was expensive first cost versus cheap running. If the first cost could not be made reasonably close to gas engines, the diesel would simply not be bought.

Specific problems awaited the tractor diesel engine. An injection system capable of governing the engine while still being cheap to produce was the major prob-

This is probably the first I-30 industrial tractor. The I-30 was based on the W-30, although the first one of the assembly line was an I-30. Tractor Q-1825 weighed 5,800 pounds and was photographed November 20, 1931. *State Historical Society of Wisconsin*

The 10-20 orchard tractors had made a market niche, so IH designed an orchard version of the W-30. While the fenders are correct, the air cleaner pipe would not have been very popular in the trees. The manifold heat controls had undergone some changes and a swinging drawbar has been added according to the IH caption. Tractor Q-1806 weighed 4,460 pounds. The photo was taken October 15, 1931. *State Historical Society of Wisconsin*

The "15-30 Gear Drive Tractor (Increased Power)." IH applied the same ideas to the 15-30 that it was trying with the 10-20, Regular, and F-30. This 15-30 has the new manifold, tall air cleaner, and except for some different piping and hose arrangements at the front of the engine, looks like a larger carbon copy of the 10-20 during this phase. For unknown reasons, IH decided to go with a six-cylinder engine for this tractor. Tractor Q-1439 weighed 6,380 pounds and was photographed December 29, 1930. *State Historical Society of Wisconsin*

lem. Another problem was ignition of the fuel. Pure compression is usually not enough to ignite the fuel-air mixture. Precombustion chambers that are hotter than the rest of the cylinder were used by International Harvester as early as 1919 in a 10-horsepower single-cylinder engine. Diesel development seems to have been put on the back burner during the early to mid-1920s, although some work with injectors was done.

International diesel development really took off with the examination of a Junkers diesel engine in 1927. The injection system featured an arm that controlled how much fuel each injector received by varying the stroke of the individual unit injection pumps. (The first IH production injection systems used individual pumps for each

This is one of the first photographs of the W-40 with a six-cylinder engine. The engine seems to have come from IH's truck division. The reason for the change from an increased power four-cylinder to a new six-cylinder is unknown, but was probably done to increase reliability by lowering the power generated per cylinder. Tractor Q-1952 weighed 5,500 pounds. The photo was taken March 4, 1932. *State Historical Society of Wisconsin*

IH had experimented with diesel engines since 1917, but the first recorded IH diesel tractor was not built until late 1928. The installation appears to have been fairly simple, with a covering added to the right side hood louvers, which probably provided clearance or access to the injection pump. This is probably from the second series of three 15-30 experimental diesel tractors built, seen on August 9, 1929. Tractor Q-738 weighed 6,840 pounds. *State Historical Society of Wisconsin*

cylinder until the introduction of the famous "Type A" injection pump with either one or two pumps in 1941.) For the first time IH had an engine that could run consistently. Now all the company had to do was start it and figure out how to produce it cheaply enough to put it into a tractor.

The first new International Junkers-influenced engine was ready for testing in September 1928. After testing in a test stand, the 4 1/2x6 engine was installed in a 15-30 chassis and shipped for testing to Phoenix, Arizona. There, the tractor operated nearly continuously until March 1930, with good results, although some IH photos show a problem with carbon buildup on the piston. During the testing, IH completed three more engines, this time 4 3/4x6 1/4 with a 14.5:1 compression ratio. One source stated that these engines put out "81.3 psi maximum BMEP at 1,100 rpm." These engines also went into tractors for testing.

International also built a single-cylinder test engine in 1930 to examine various fuel-injection and combustion systems. This engine was fitted with windows in the combustion chamber so that a high-speed camera could be used to watch the combustion process, one of the first uses of this technique. IH developed the more familiar single-piece, multiple-pump injection unit at about this time.

Diesel Starting Systems

The Gas Power Engineering Department experimented with several starting systems. Many of the experiments followed the path of many others: a starting engine (or "Pony" engine) that could be started with a hand crank and then clutched to the larger engine. IH had much experience with these starting engines back in the teens for starting the large gas engines. International experimented with both single-cylinder gas and diesel starting engines for the diesels, but never put the system into production.

Another system IH experimented with was a furnace system. An auxiliary furnace was attached to the tractor so that heat could be circulated through the engine to assist starting. The draft for the furnace was supplied by the farmer hand cranking it, presumably for a rather long period. Prototypes of this system were built and tested, but again never put into production.

The third, and most unusual, system was that of starting on gasoline. International did not originate this system. It was Waukesha Motors, a familiar partner of IH, which first put the system into use. Waukesha had developed a system using a hand lever that closed the main air intake and opened the valve leading through a carburetor. This simultaneously lowered compression (making the tractor easier to crank) while introducing a gasoline charge to the cylinder where it was ignited by a spark plug. The same hand lever also closed the circuit to the ignition system. When the motor had warmed up, the

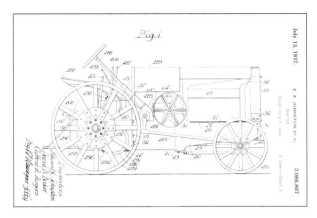

The steering clutch tractor patent. Not only did this tractor have a normal steering wheel and turning brakes, it also had two steering clutches so that power could be controlled to either driving wheel. The clutch housings can be seen as part 230. The steering levers are 206 and 287. The tractor was designed to be easily convertible to a crawler tractor. Production of crawlers and wheel tractors contain the maximum number of common parts. *U.S. Patent*

lever was thrown back and the engine operated as a diesel. IH improved this system by devising gearing by which the switch from gasoline to diesel operation was done automatically after a certain number of revolutions, usually about 900.

Improved Power Steam: The Locomotor

The success of the 15-30 and 10-20 programs killed the steam tractor idea in the early 1920s after several of the experimental tractors had been built and tested. International Harvester probably had a tractor design that would work in the field, but manufacturing costs and life of the tractor would probably have presented problems. Harvester engineers kept track of steam development and

Under the hood of an IH diesel tractor. This engine was 4 1/2x6 1/2, with an unknown starting system. The unit construction system enabled IH to put in experimental engines with a minimum of rework to the chassis. Tractor Q-896 weighed 6,920 pounds and was photographed February 14, 1930.

The W-40 shown here had a 4 1/2x6 1/2 engine. Diesel engine development involved settling many questions outside of the engine itself, including starting systems and the difficult task of producing the engine at a price that the farmer could afford at a quality that would stand up over time. IH eventually air conditioned the precision manufacturing areas of Milwaukee Works in order to gain greater machining accuracy. This is tractor Q-1645, seen September 25, 1931.

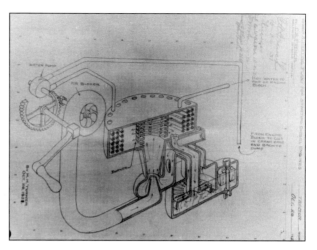

One of the diesel starting systems examined by IH included a furnace designed by C. A. French, who had been heavily involved with the design of the fuel burners for the IH steam tractors. This system involved a hand-cranked air blower and water pump. The blower forced a draught through the furnace, which warmed water that was then pumped through the engine block and through a coil in the crankcase in order to warm the engine to assist starting. Apparently this drawing was a preliminary stage to a patent, probably done more for protection than for actual application purposes. *State Historical Society of Wisconsin*

built a few other steam machines, most notably two steam rail passenger cars in the late 1920s and early 1930s that used many of the ideas of the steam tractor, including a flash boiler and two V-8 Uniflow engines. The main IH engineers were Charles A. French (expert on "blue flame" or sootless combustion), Louis Goit, and Gustaf

Engstrom. All of these engineers had worked on IH's steam tractor project. French and Goit had left IH in the early 1920s and went to the Endurance Car Company of Los Angeles, California, where they worked briefly on a steam car project that used a burner and boiler similar to that of the IH steam tractor. French also worked with the

A swash plate engine that was experimented with by IH. The pistons in a swash plate engine are aligned along the driveshaft of the engine. The connecting rods end against a warped plate. As the pistons go up and down, the connecting rods push along the slopes of the warp, forcing the plate to rotate. This idea was used in a few engine designs in the 1930s and 1940s, including some by British aircraft engine companies that saw production. The horsepower and proposed use of this IH engine is unknown. This engine was photographed January 10, 1935. *State Historical Society of Wisconsin*

IH's venture into the railroad self-propelled passenger car business was powered by a Uniflow V-8 steam engine as shown here. Two of these were located under the car, directly connected with the bevel gearing on the truck. A unique application for a unique engine. The photo is dated February 4, 1929. *State Historical Society of Wisconsin*

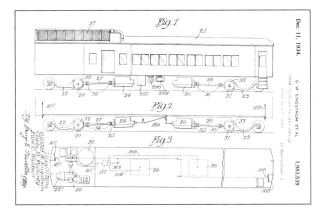

The side view of the Locomotor. Figure 1 is the side view as the Locomotor would be seen, Figure 2 gives better detail of the engine control linkages, while Figure 3 is a top view showing internal details. The equipment shown by the dotted lines is actually below the floor of the passenger compartment. *U.S. Patent*

Coats Steam Car company of Columbus, Ohio, in the early 1920s.

The Locomotor was a complicated piece of machinery when compared to tractors of the time. Passengers had to be kept comfortable and safe, while electricity had to be provided for lighting the passenger compartment and headlights. The engines of the Locomotor had to have enough steam to provide enough power to not only rapidly accelerate the Locomotor up to top speed, but also to do the same for at least two more cars pulled behind. A system had to be in place to control the two engines simultaneously. And finally, air brakes had to be provided to not only the Locomotor but also for the other cars.

There were many reasons why the Locomotor could have been an economic success. The rise of the automobile had reduced traffic on many short or out-of-the-way rail lines. The steam locomotives of that day had some definite faults, especially in service on low-density passenger routes. The steam locomotives were expensive to run, produced large amounts of smoke, and took time to build up pressure when starting with a cold boiler. The Locomotor, on the other hand, was more fuel efficient, could build up pressure quickly from cold, and produced much less smoke. Lowering the expenses on lines with small amounts of traffic could change a loss to a profit. Railroads had been building or buying railcars equipped with internal-combustion engines, but often the engines didn't have enough horsepower or were mechanically unreliable.

The Locomotor's boiler was automatically fired, eliminating the need for a fireman. Boiler pressures ran between 600 and 650 psi, about the same or slightly higher than the steam tractors. The steam was superheated about 250 degrees. However, the boiler horsepower of the steam tractor was about 50, while in the Locomotor, boiler horsepower ran at about 500, run-

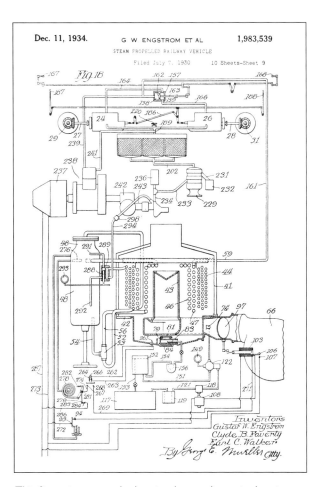

This figure is not a scale drawing, but a schematic showing the different equipment and appliances involved in the mechanical system of the Locomotor. It was a very complex system. *U.S. Patent*

ning two 225-horsepower engines and several boiler and train auxiliary units.

The boiler (or steam generator) was arranged around the burner. The tubing nearest the flame was "Enduro" metal, a chrome-nickel heat-treated steel. The nearest coils were filled with "wet" or "saturated" steam, which is low-temperature steam, mainly to keep the coils cooler. The near coils had about 85 feet worth of Enduro tubing. The main burner area reached a temperature of 3,500 degrees Fahrenheit, although the burner itself was kept much cooler by air flow. Starting of the burner was by electric grid.

The next layer of tubes contained the "dry" or high-temperature steam that would be going next to the engines. This tubing, also called the superheater section, had about 75 feet of Enduro tubing. The tubes farthest away from the oil burners would contain water, which was heated by the lower temperature exhaust fumes. Keeping the cooler tubes farther out also acted to insulate the boiler, keeping the outer shell cool. Total amount of tubing was about 1,000 feet, with about 385 feet being Enduro metal. The remainder was low-carbon steel seamless tubing.

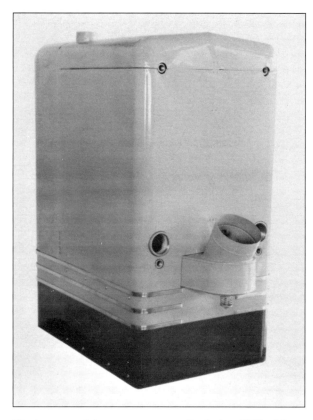

The last gasp of the steam program. By the late 1930s, IH was growing increasingly interested in products for the home, as an outgrowth of the refrigeration and cooling program, which itself was an outgrowth of the dairy product line. IH designed this furnace, which used low-pressure steam. The serial number of the furnace was Q-3655; weight was 1,520 pounds. The photo is dated November 4, 1937. *State Historical Society of Wisconsin*

The boiler system was controlled by steam pressure. Control in a flash boiler is difficult. In a normal steam boiler, there is plenty of excess capacity in order to smooth out variations in steam demand from the engines. In a flash boiler, the steam is in one large continuous tube of much smaller volume than a traditional boiler. Changes in demand have to be met with changes in fuel flow, water flow, and temperature. The Locomotor used a steam storage drum to provide some of the advantages of the traditional boiler while still maintaining a flash system. The storage drum maintained a small supply of steam in the upper half of the drum, which provided a reserve in case of a short-term rise in demand in steam, such as a small hill. If the demand was of longer term, the pressure would drop. The storage drum's lower half was filled with water just under the boiling point at the operating pressure. Water under pressure has a very high boiling point. When the pressure dropped, so did the boiling point of the water, which then turned to steam needed by the engines. The same drop in pressure also activated or turned up the heat in the burner. The storage tank also enabled a gauge to be placed on the system. The gauge measured the total amount of water in the system, critical in a condensing system where the water is constantly circulated.

As mentioned above, the Locomotor had a complicated set of auxiliary devices. These devices included a gas-electric powerplant (built by Kohler Company) to provide power to the burner ignition grid, a fuel pump, the burner air blower, a steam-oil separator to remove the cylinder oil from the condensed steam so that it could be run through the boiler again, a steam-driven electric generator, condenser fans, a 32-volt lighting system, and the compressor for the air brakes. The entire auxiliary system was controlled by a single drum-type controller in the control cab. The controller, similar to the type used for engine room signaling in ships, was arranged so that the various positions corresponded with the start-up procedure. Position 1 started the gas engine-powered generator; position 2 started the electric grid that ignited the boiler burner. Position 3 actually started the fuel pump, supplying fuel to the burner vaporizer pots, which heated the fuel oil to the point where it would form a fine, ignitable mist. The fourth position started the burner blower, providing forced air to the burner. Another butterfly valve actually opened the vaporizer pots, starting ignition. After ignition, it took only one minute for enough steam to be generated to operate the four-cylinder Uniflow auxiliary engine that drove the main electric generator. After the steam was raised, position 5 started the oil separator, while in position 6 the electrical load was transferred from the gas engine generator to the main generator. At position 8, the condenser fans on the roof are started. If starting absolutely cold, it took the Locomotor 20 minutes to raise enough steam to start rolling under its own power. If the Locomotor had only been sitting less than an hour, it only took four minutes to get up a full head of steam.

Once steam had been raised, it was piped to the two main engines. Each Locomotor had two 225-horsepower engines. The engine design varied greatly between the two Locomotors. On the first Locomotor, a straight-eight design was used, while on the second Locomotor, a 135-degree V-8 was used. The engines had intake cutoff and exhaust controlled by a special camshaft assembly. For each cylinder, the camshaft had several different cams. The camshaft could be moved lengthwise to change which cams were controlling the valves. The engine was reversed using the same process.

The power produced by the engines was then sent to a differential and transferred to the wheels. On each two-axle truck assembly, one axle was powered. A special torque beam ran between the powered and unpowered axles to control rolling.

The car bodies were produced by the Pullman Company, with assembly of all components probably done by Ryan Car Company, with the completed Locomotors going to the Chicago, Milwaukee, St. Paul and Pacific Railroad (the Milwaukee Road). The cars seem to have

The vegetable-powered tractor! More seriously, this is a gas generator connected to a W-30. Gas generators converted material, such as wood or other vegetable matter (or coal) directly to a combustible gas. Gas generators themselves were an older idea (IH had manufactured them before 1910) that made a return in Europe (and a few places in the United States) during the fuel shortages of World War II. This attachment was photographed August 11, 1935.

with the most notable use being certain British aircraft engines of the later stages of World War II.

Another unusual development was the vegetable-powered tractor. An attachment was designed that would take vegetable matter and process the material into producer gas, which then would be burned by the tractor. IH had designed producer gas systems in the early years of stationary engine production (some patents granted to Milwaukee-based engineers date to 1908), so they were familiar with the process. The combination of the Depression and the Chemurgic Movement (where more industrial uses for farm products were experimented with to increase demand and income for farmers) probably prompted International Harvester to see if farmers could produce their own tractor fuel as they had in the days of horses. Devices were built and tested, but little seems to have come of the project.

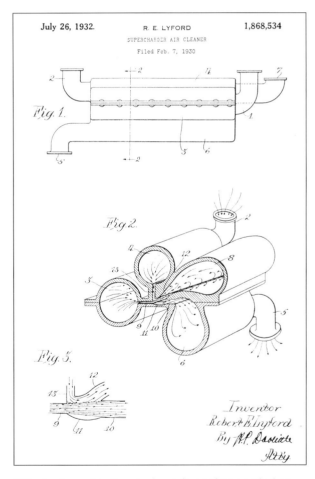

This simple combination air cleaner/supercharger relied on air flow and inertia to do its job, somewhat similar to a carburetor. The dust and grit in the air would be passed to the exhaust gasses as shown by Figure 3. The intake air could change direction quickly, while the dust would go in a straight line into the exhaust. The high-speed exhaust, meanwhile, would impart some of that speed to the intake air, forcing it into the cylinders. *U.S. Patent*

been successful initially. During initial testing, David B. Baker took a Locomotor up to 86 miles per hour between Chicago and Milwaukee on the Milwaukee Road, a track known for high-speed operation. The first Locomotor ran a difficult route between Milwaukee and Mineral Point, Wisconsin, very successfully. One article mentions that up to six Locomotors were on order, although it does not say which railroads had ordered them. However, the IH-driven steam cars were beset with mechanical difficulties and development was not continued. At first, the roof-mounted steam condensers were actually too effective, causing water to back up and plug the system. This problem was eventually solved. The advanced flash boiler was too advanced for the materials then in existence, resulting in early and frequent boiler failures. Tubes in flash boilers are often run at extremely high temperatures in close proximity to combustion flames and gases. The tubes probably eroded quickly, causing the eventual termination of the program.

Alternative Engine Development

Wobble plate engines were experimented with at least briefly by IH engineers in the early thirties, proving that nothing was sacred. Wobble plate engines have no crankshaft, but instead have a curved rotating plate that the pistons ride up and down on. Wobble plate engines were going through a period of great interest during this time due to their lack of a crankshaft and potential large horsepower in a small engine. Although IH's engine apparently never got too far beyond the concept stage, other companies did succeed in building these engines,

Chapter 7

12 Series Tractors

Although the F-12 is often thought of by many as simply the small end of the F-series, the archival photos and the little written information that survives tell a very different story: that of the completely separate 12 series.

An important measure of success for IH's tractor marketing was the number of horses remaining on American farms. Throughout the twenties and thirties, many horses remained on American farms, especially the smaller farms. For some farmers, even the Farmall and the 10-20 were too big and expensive. When the horses finally died or were no longer cost-effective, many of these farmers bought the cheapest, smallest tractor available—most likely a Fordson, but possibly one of the garden tractors that were now coming into the market. International would debate the merits of producing a garden tractor until the advent of the Cub and ultimately the Cadet line.

However, there were still plenty of farmers who could be sold a tractor in the Fordson range, or at least a tractor smaller and cheaper than the 10-20 and F-20.

Letters from one dealer appear to have had a special impact, as they were given a prominent place in the little information that has survived from that era in International's files. The H. S. Allis Company, Inc., from Providence, Rhode Island, handled many different lines of equipment for the farm, construction, and industrial markets. The dealer noted the tight confines of construction sites, the small hallways of older facto-

ries, and the huge number of smaller farms that existed in New England. In one letter H. S. Allis wrote that many of the smaller Holt crawler tractors had been sold to contractors and road constructors, and that contractors were more interested in the smaller units. He also noted the coming of the new Fordsons, which by then were coming from Ireland. He felt that IH could sell many tractors if the company could make a smaller tractor with the features of IH's regular lines of equipment.

Allis had already sold 2,000 Fordsons. This dealer estimated that the demand for a new International Harvester small tractor would be about double the demand for the 10-20. Such numbers apparently got IH management's attention. He also noted that the tractor should be made to accept rubber tires and mounted equipment. In addition, Allis indicated that the tractor should be able to have a crawler attachment added to it and he mentioned the Wehr attachment as being serviceable and small.

Very little other documentation exists for the small tractor program, and the photo information may

The "Special Industrial Narrow-Tread Tractor." Not much is known about this tractor, but it is believed that this was the first step toward the 12 series. The body looks like a regular 10-20, but after that it looks like a lot of new parts. Tractor Q-131 was photographed February 13, 1929. *State Historical Society of Wisconsin*

The "Small Industrial Tractor" was a more refined attempt to build a small tractor. Although engine size is not known, it looks as if the frame and other parts have been scaled down to build a proportionate tractor. Tractor Q-995 weighed 3,850 pounds, photographed June 9, 1930. *State Historical Society of Wisconsin*

The "Model 15 Industrial Tractor Using T-20 Main Frame." The T-20 crawler tractor was originally titled the T-15. This was a clutch steer tractor, basically replacing the crawler assembly with rear wheels and then adding a free pivoting front wheel. Result: a very short steering tractor that shared many parts with other IH equipment, resulting in reduced cost of production. A neat idea, but even with the shared parts, it was almost certainly an expensive tractor to build. *State Historical Society of Wisconsin*

be incomplete. Apparently IH's Gas Power Engineering Department had in mind a series of tractors that would meet all three of the dealer's requirements, originally described as the 10 series (which would fit in with the 20 and 30 series tractors then in development). A crawler, an industrial, and a Farmall tractor were experimented with. At some point power increased to the point that the series became the 15 series, probably very early in development.

Crawler and industrial tractor development seems to have taken primary importance. The crawler tractor eventually received a higher horsepower rating and became the T-20 after a few T-15s were produced. The industrial tractor seems to have had the most interesting

This is the Sloniger tractor, an obvious attempt at a shop-mule tractor. A very different steering system, probably with clutches, was used. Sloniger was an engineer at Milwaukee Works. It's possible that this tractor was designed for internal use, as IH used hundreds of shop trucks, industrial tractors, lifts, and other machines in its own factories. There is no Q number or weight given for this tractor, which adds credibility to the internal use idea. *State Historical Society of Wisconsin*

In the photo caption for this photo, F-10 is crossed off and F-15 substituted. The earliest ancestor of the F-12, this tractor was the first Farmall to do away with the heavy, expensive side housings and final drives, instead using a large-diameter rear wheel to gain clearance over the crop. Tractor Q-1938 weighed 1,997 pounds and was photographed February 22, 1932. *State Historical Society of Wisconsin*

The F-10 Farmall, although this is tractor Q-1938, which was originally called an F-10, then F-15. The steel channel frame no longer runs the length of the tractor, but instead the transmission casing forms the rear frame, saving weight. The seat support of this tractor will be changed, as will the general design of the rear transmission casing. Tractor Q-1938 weighed 1,997 pounds and was photographed March 17, 1932. *State Historical Society of Wisconsin*

The IH caption for this photo originally read, "Fenders for Farmall Tractor F-15," but the F-15 was crossed out and replaced by "F-12." The wheels have been made heavier, and part of the steering pedestal has been changed. The fenders, a legal requirement in at least one state (Wisconsin), do look out of place on this tractor. Tractor Q-2078 was photographed May 26, 1932. The fenders weighed 68 pounds for the set. *State Historical Society of Wisconsin*

development. The first small industrial tractors were clearly based off the 10-20, and the first one seems to be a 10-20 with smaller wheels and shorter wheelbase. IH then built a similar tractor with a smaller engine.

International also experimented with a totally different design from its usual tractor designs, the I-15 having more in common with the crawler than the wheel tractors. A tractor was devised with the main frame and steering clutches of the T-20 crawler, but with back wheels instead of treads and a free-swiveling narrow front, turning being through the rear clutches and brakes. This resulted in extremely short turning radiuses. Interestingly enough, Farmall development apparently does not begin until after this point, suggesting that the Farmall member of this series was to have the same features. Given the appearance of this tractor, a Farmall version certainly would seem to be natural. At about this point, however, tractors with conventional steering were designed, with the Farmall version coming first. IH continued with the idea of steering clutch tractors. A patent for the idea was applied for in 1934. The patent application states that the main idea is to provide a form of tractor construction that could be used for both crawler and wheel tractors, which of course would remove the need for separate tractor designs and reduce expenses. However, this tractor was not developed further, and tractor development along conventional design commenced.

Sloniger Tractor

There was interest in a smaller industrial tractor, and IH's Milwaukee Works apparently built one known as the Sloniger tractor, presumably after the engineer who devised the configuration. The tractor (if it can be called that—it was more of a shop mule) was extremely compact. This design was apparently not pursued much further.

F-12 Prototypes

As mentioned above, the small Farmall program started off with the tractor being called an F-10 or F-15. Design probably started in late 1930 or early 1931, and by early 1932, experimental tractors were being tested. The Farmall-12 program occurred at the same time as the efforts to design the improved power line of tractors and a time of serious truck engineering. In addition, several other projects were also taking engineering time. This may be one reason for the F-12 being produced for a short time with an engine bought from an outside supplier, a distinction shared with the Motor Cultivator and early 15-30s. Harvester was developing a 3x4 engine during the time the F-12 was being developed, but may have been delayed by other work. However, there are several other possible explanations as to why the F-12 was powered by a Waukesha BD-1 engine.

This engine itself has an interesting history, beginning in 1929. Harvester had for many years purchased small engines from outside suppliers for binders, harvester-threshers, hay presses, and potato diggers, producing larger engines itself. In 1929 IH brought in a team of engineers from Fuller & Johnson, a prominent engine builder from Madison, Wisconsin. F & J had fallen on hard times, struggling with the costs of developing a new engine line to replace the traditional hopper-cooled single-cylinder engines. By 1929, the com-

Golf course wheels on an F-12. For some reason the provision of golf course wheels on F-12s caused debate within IH, with the decision to manufacture being made and then changed several times. The wheels weighed 768 pounds, resulting in a total tractor weight of 2,585 pounds. The photo was taken May 10, 1935. *State Historical Society of Wisconsin*

This photo is titled, "Fenders on O-12 Tractor," although the fuel tank shows the "I-12" decal—an indication of how close these tractors were developmentally! The experimental fenders were Q-2883 and weighed 208 pounds per set and were photographed March 25, 1935. *State Historical Society of Wisconsin*

pany was in serious trouble, but had new high-speed radiator-cooled engines. Why Harvester turned to F & J out of the many other companies producing these engines is hard to determine. Harvester did have a policy of having as many sources for a particular purchased item as possible in order to avoid getting "stuck" with high prices or bad quality, so perhaps F & J was brought on to have one more source of engines. Another possibility is more personal: Alexander Legge, president of IH, grew up on the farm owned by Morris Fuller, one of the founders of F & J. Fuller gave Legge's father a job as manager of his Nebraska ranch after the older Legge had suffered a severe financial loss, through no fault of his own. Perhaps Alexander Legge tried to give Fuller & Johnson a helping hand. Legge's biography specifically mentions that Legge had a habit of assisting other companies through tough times by helping them modernize, although the biography gives a different example and no records could be found to confirm this idea.

Whatever the reason, F & J's engineers designed and built a sample engine that IH found acceptable. By this time, F & J was in such bad shape that it could not

Here is the first Fairway 12, titled "Fairway 'Golf Course Wheel' Attachments on I-12 Tractor." Wide tires provided enough flotation for the tractor so it would not damage golf greens but at the same time rolled the greens. It is a rare and desirable collector's item today. Tractor Q-2470 weighed 3,010 pounds and was photographed August 2, 1933. *State Historical Society of Wisconsin*

afford new machinery to economically produce the engine in the quantities that IH would soon be buying. Waukesha Engines had acquired an interest in F & J in 1925 after the Johnson family had decided it was time to get out of F & J and concentrate on the main family business, Gisholt, which manufactured machine tools, notably turret lathes. Waukesha had been cooperating with F & J closely, pooling purchases of raw materials for larger discounts and contracting work out to F & J

(and vice versa). Waukesha had more modern tools, so it took over the tooling and began production of the first 500 engines for IH. First use was in IH harvester-threshers. Later, IH would have an emergency demand for an auxiliary engine for hay presses for California. The F & J engine proved extremely handy in solving this problem. Shortly thereafter, in 1931, F & J declared bankruptcy and Waukesha purchased the entire BD program, including tooling, drawings, and contracts, for $10,000, a price that also included two other slightly different engines that used most of the parts of the BD.

Waukesha was presented with a new market for the BD engine that must have convinced them they had gotten a bargain. The first F-12s used the Waukesha engine (now renamed the BD-1, possibly after some modifications were made). Why these first tractors were produced with the BD-1 engine is, as mentioned above, a mystery. Whether it was due to a lack of engineering time or uncertainty about the tractor (the Waukesha-engined tractors could have been preproduction tractors intended to test the small tractor in the field in the hands of users), the Waukesha was replaced by an IH-engineered and produced overhead valve 3x4 engine on May 23, 1933, on tractor FS-3840. The change had been scheduled to take place on FS-2527 but was apparently delayed for a short time.

The engine was not the only innovation present in the F-12. The F-12 was also the Farmall to do away with the heavy final drives and side housings. As men-

The Steam Tractor and the F-12/12 Series

Although the steam tractor was never produced by International Harvester, it played a role in IH's steam Locomotor project. It may have also played an important role in IH's tractor development in the 1930s. The steam tractor used a steel channel forward frame that was attached to the transmission case that formed the rear frame, just as the F-12 and other 12 series tractors used.

The 12 series tractors were engineered by IH to significantly reduce the size of a particular horsepower of tractor, just as the steam tractor was. Compare the statement, "The main purpose of the invention is to provide a light, simple structure for a tractor of the type stated . . . and one in which the mechanisms of the power transmission line and principal parts of the brake applying means are contained within the structure of the tractor body, thus affording a compact body structure of the tractor body, thus affording a compact body" (Patent 2,031,317, regarding the layout of the F-12) with, "The

object of my invention is, therefore, to provide a steam tractor that may be made in small sizes of equal and greater power and of less first cost and operating expense than the corresponding sizes of gas engine tractors now on the market. It is also an object of my invention to arrange and correlate the several parts of a steam propelled tractor so as to provide a neat, compact and efficient power plant and tractor" (steam tractor Patent 1,685,279).

Both the 12 series and the steam tractor used transverse transmission, which was unusual in IH engineering practice. The question must be asked—when IH was trying to build a small gas tractor in the early 1930s, did the company look back on the small steam tractor of the 1920s and recycle some of the ideas? While we may never know the answer, there seems to be more than a coincidental resemblance.

The experiment in this photo is probably the start of the F-14. Titled, "Parts for Bringing the F-12 Up to Date," this shows changes in the manifold and an interesting apparatus above the valve cover that probably was some sort of vaporizing device for heavy fuels such as distillate or kerosene. The total weight of the changes was 38 pounds. Photo taken July 31, 1936. *State Historical Society of Wisconsin*

Titled, "Modifications on F-12 Tractor," this photo shows further modifications to the F-12's manifolds, including what looks like a very unusual muffler and an under-the-seat lift system. Tractor Q-3406 weighed 3,495 pounds and was photographed November 17, 1936. *State Historical Society of Wisconsin*

tioned in the prior chapter, J. I. Case has pioneered the row-crop tractor without side housings, instead using larger wheels and straight axles. This had many advantages, including lower weight, cost, and simplicity. The design also allowed a greater range of available treads for the rear wheels. To change tread, all a farmer had to do was to loosen nuts on the wheel and slide the wheel along the rear axle. On the other Farmalls, the heavy wheels and wheel centers had to be taken off and flipped. The F-12 also had a more advanced frame. On the F-20 and F-30, the girder frame extends to the end of the tractor. On the F-12, the transmission case forms the rear part of the frame, again saving weight and money.

The Gas Power Engineering Department patented several other features of the F-12, including the general layout and the transmission. IH deviated from its normal engineering practice, most notably in the transmission, in order to produce a smaller tractor. The first 25 preproduction F-12s were authorized for production August 10, 1932. Regular production was authorized December 5, 1932, with the first regular production tractor, FS-526, coming off the line January 11, 1933.

Sales were brisk for the F-12. However, questions were soon being asked about the size of the tractor. Some people within IH thought that the tractor had been made too large. The two-plow capacity in good soil was the same as the previous Farmall Regular. Could a yet smaller Farmall have found an even bigger market? The kick against the F-12 was that while it had a two-row cultivator and two-bottom lister, it was only rated as a one-plow tractor and was expensive for that type of tractor. When Allis-Chalmers brought out the first "mini" tractor, the B, these people would be heard from again.

International Harvester provided golf course wheels for the F-12. According to the decision authorizing them, the golf wheels were made in "very limited quantity" by hand.

Other 12 Series Prototypes

International proceeded with a series of wheel tractors based on the F-12. The first was the I-12, about which little if any information can be found about its development. The O-12 was another F-12 derivative, which differed from the I-12 originally only in brakes and drawbar. Both the I-12 and O-12 had 2,000 rpm engines and variable speed governors, as opposed to the F-12's 1,400 rpm. Of course, both tractors had different frames, steering, and other changes to convert them to standard tractors. The O-12 was authorized for production December 4, 1933. The Fairway 12 was an O-12 with steel wheels instead of rubber, but with the I-12's foot brakes. The Fairway was authorized for production May 1, 1934.

The W-12 was developed from the O-12, but with the F-12 1,400-rpm engine and a slower operation than the I-12 and O-12. This tractor was authorized for production December 4, 1933, and was in production February 9, 1934, on tractor WS-503. A new version of the W-12, with a 1,700 rpm engine, was released for production October 23, 1934.

F-14 Development

By 1935, only two years after entering the market, IH's engineers were already attempting to bring the F-12 up to date. Ironically, these measures added even more horsepower by changing small details such as the mani-

89

fold. The F-14 was authorized for production December 23, 1937, and included an increase of rpm to 1,650 from 1,400, a new cylinder head to increase compression, new bevel gears in the transmission, a new seat giving an improved riding position while allowing the operator to stand, a change in the position of the clutch pedal and hand brake levers to allow for the new operator position, and new power take-off gear were included (IH labeled the original F-14 as "F-12-General Improvements"). The goal of the changes was a more comfortable operator on a more powerful tractor. The first F-14 was built January 27, 1938, tractor FS-124000. The last F-12 built was the same date, but as tractor FS-123943. The name was officially changed February 14, 1938. The official change for

Development of the "F-12" continued right into 1939, despite the renaming of the production tractor to F-14 and the massive program to design and build a replacement. This particular development, however, potentially had applications to the new tractor. This is a two-cylinder diesel engine for the F-12. The presence of the large blower indicates that this engine was probably a two-cycle, two-cylinder engine. A novel idea, well ahead of its time, but it probably ran into the problem of what farmers wanted in a small tractor: a small price tag. Engine Q-3881 weighed 729 pounds and was photographed February 16, 1939. *State Historical Society of Wisconsin*

The International Harvester "Power Driven Lawn Mower." The engine looks like a Briggs, although it is believed that IH experimented with small air-cooled engines several times. IH was looking into ways to supply its dealers with product lines that were being purchased from other companies. Dealers wanted the wider product line in order to maintain cash flow and use overhead in the notoriously seasonal farm equipment industry. A product line that appealed to non-farmers was also a benefit in the 1930s. Still, nothing would come of this until the Cub in 1947, and ultimately the Cub Cadet line of the early 1960s. Mower Q-2088 weighed 290 pounds (not a lightweight machine!) and was photographed June 10, 1932. *State Historical Society of Wisconsin*

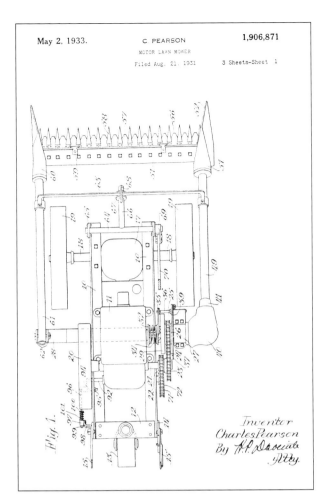

The patent for the IH lawnmower used an LA-type hopper-cooled engine (the hole in the hopper is labeled 17). The patent drawing is different from the photographs mainly in the drive area. The photos have a bulge on the right-hand side of the sickle to enclose the sickle drive mechanism. The patent does not. Also different is the wheel drive mechanism and the entire control layout. *U.S. Patent*

I-14, W-14, and Fairway-14 tractors came April 11, 1938, although they had been in production since March 22. The new F-14 engine was used with the older, higher rpm for each tractor.

The situation for the O-14 was a little different. One hundred O-14s were made to meet "immediate requirements" with the eventual building of 300 more O-14s with the new engine, a new low seat support with a large implement-type seat rearranged to allow the operator to stand, new citrus orchard fenders, a new dropped platform, swinging drawbar adjustable to three positions, a spring-loaded clutch for the first 100 and a regular modified O-12 over-center clutch for the 300 regular production tractors, countershaft, foot-operated brake with strengthened parts, a shorter steering wheel shaft, a rear exhaust pipe, repositioning of the governor control level along with other control repositioning, and a change in wheel equipment. The first of the 100 pre-production tractors was OS-3499 (built April 5, 1938), and the first regular production tractor was OS-3610 (built June 27, 1938), indicating that more than 100 preproduction tractors were produced.

Fenceline Mowers and Air-Cooled Engines

International Harvester management was aware) that even the F-12 wouldn't completely erase the horse from the American farm. They also had followed with interest the rise of small, walk-behind powered implements used in yards and gardens and a specialized implement referred to as a fenceline mower. Fenceline mowers were basically a small sickle bar with attached motor, wheels, and handle. IH produced several prototypes starting in the early 1930s and never really gave up on the idea until sometime after the Cub came out in 1947. The idea may even have lasted until the Cub Cadet era of the 1960s.

Although the patent for the IH Fenceline Mower states, "The principal object of the invention is to provide a simple, compact, power operated mower," it is clear from the patent that the IH mower, as it then existed, was anything but simple and compact. It had a unique, complicated differential, a somewhat dubious feature on a hand-guided lawnmower. The construction of the mower in both the photograph and the patent appears to be extremely heavy. It also featured a reverse gear, which also probably increased cost and weight. If the cutter bar struck an obstruction or became overloaded, a safety clutch disconnected the power to the bar. The cutter bar also could be lifted to go over obstructions and featured a transport position! These features probably go a long way in explaining why the mower never was produced.

Air-cooled motors were also experimented with in the 1930s. The Briggs and Stratton and other makes cut heavily into IH's small water-cooled engine line. The popularity of the air-cooled engine also hinted at a vast new market for Harvester. While air-cooled engines were designed, built, and tested, IH never put them into production due to the competition already in place.

International apparently did some research into rototillers in the mid-1940s. Ed Johnston's last patent was for a unique rototiller using rotating disk blades and powered by a very unusual engine. The engine may have been the fantasy of some patent draftsman (the engine had no bearing on the actual patent). In the patent description, it states that the rototiller should also be usable when mounted on a tractor, which may have been the real intended use.

Chapter 8

Letter Series Development

By the mid-1930s, IH's wheel tractor line was starting to become dated. Except for the diesel tractors, most of the designs were now 15 years old. The Farmalls had never been considered an attractive tractor by IH or most farmers and implement men. Newer technology was coming into the tractor business, which made IH's tractors seem even more obsolete. The use of more modern materials and methods by Allis-Chalmers, resulting in lower tractor weights and costs, along with the use of pneumatic tires were probably the most important changes of the mid-thirties. Streamlining started to affect tractor design a little later in the decade, while hydraulics were also starting to take hold of the American farmer. All of these combined spelled a need for new tractors.

The Lightweight 20 Series

In May 1935, Gas Power Engineering produced a "Light Model F-20 Farmall Tractor" and in March 1936, a "O-20 Orchard Tractor." These tractors both used a 3 3/4x5 engine. The tractors used the same style of transmission case and frame as the F-12 and O-12, and on the F-20, omitted the final drives and side housings. It would be fair to call these tractors scaled-up versions of the 12 Series tractors. By March 1936, the Lightweight F-20 had become the F-21 with a revised front bolster and a mechanical implement lift under the driver's seat. There are records that indicate that IH sold nine of these tractors to farmers, almost certainly for testing purposes. Up to this point, the styling was still that of the production F-20, but IH was paying attention to that as well.

Six-Wheel Tractors

While the process of developing the new Farmalls was under way, attention was paid to updating the standard tractors. The first experiments beyond the W-30 and W-40 tractors were into six-wheel tractors. IH at this time was making massive efforts at designing a complete crawler tractor line for both the industrial and agricultural markets. However, there was apparently some unhappiness with the expense of crawler tractors for the farm. The need for something better than the two-wheel-drive tractor to deal with poor traction conditions was still there. So IH hit upon a compromise—the six-wheel-drive tractor, aimed right at the space between wheel tractor expense and crawler ability.

International had experimented with six-wheel

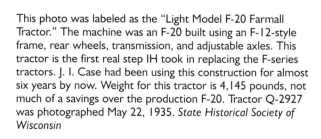

This photo was labeled as the "Light Model F-20 Farmall Tractor." The machine was an F-20 built using an F-12-style frame, rear wheels, transmission, and adjustable axles. This tractor is the first real step IH took in replacing the F-series tractors. J. I. Case had been using this construction for almost six years by now. Weight for this tractor is 4,145 pounds, not much of a savings over the production F-20. Tractor Q-2927 was photographed May 22, 1935. *State Historical Society of Wisconsin*

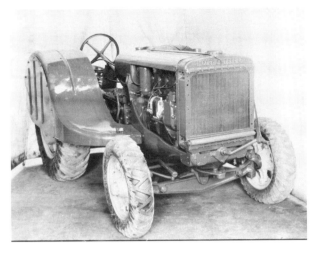

The O-20 looks a lot like a scaled-up O-12, especially in the frame area. This tractor had the 3 3/4x5 four-cylinder engine like the F-20 and the experimental Farmalls that the Gas Power Engineering Department was testing. Weight of this tractor was 4,310 pounds. Tractor Q-3191 was photographed March 4, 1936. *State Historical Society of Wisconsin*

The F-21 continued with the Light Model F-20, but was further developed. The tractor now has a lift unit under the operator's seat. The rear transmission casting has been revised, having a much smaller projection underneath. The front frame cross-member has also been revised. It is believed nine of these tractors were actually produced for farmers. Tractor Q-3126 weighed 4,100 pounds and was photographed March 1, 1936. *State Historical Society of Wisconsin*

drives before in the late teens with the IH 8-16. These experiments were probably the main reason for going with a six-wheel drive versus the more usual four-wheel drive. However, the patent drawing revealed a tractor that was more complicated than both the two-wheel-drive and crawler tractors it was meant to replace. Six different driveshafts, wheels, and, eventually, independent

IH experimented with adding a crawler drive to wheel tractors. This is the "Redesigned Semi-Crawler Attachment Having Torque Compensator Drive on W-30 Tractor." A unique feature of this design is revealed in the patent. If the crawler didn't provide enough traction, paddles could be added to the tracks for even more traction. The track attachment was Q-2182. When installed, the attachment and tractor combination weighed 6,730 pounds. This photo was taken April 19, 1935. *State Historical Society of Wisconsin*

One of the early IH six-wheel tractors, the CTW-40 was actually driven only by the rear four tires. The CTW-40 was rated as a four-plow tractor. This tractor was probably the first step in looking for a happy medium between the lower cost of a two-wheel-drive four-wheel tractor and the high traction of the full crawler tractor. Tractor Q-2670 weighed 6,260 pounds and was photographed April 1, 1935. *State Historical Society of Wisconsin*

The "6 Cylinder 3 5/8x4 1/2 Inch 6-Wheel Drive Tractor." The first of the true six-wheel-drive tractors from this period, this tractor had a suspension system that wasn't quite flexible enough for IH. However, this tractor showed enough promise for IH to continue development. Tractor Q-3077 weighed 9,750 pounds and was photographed November 21, 1935. *State Historical Society of Wisconsin*

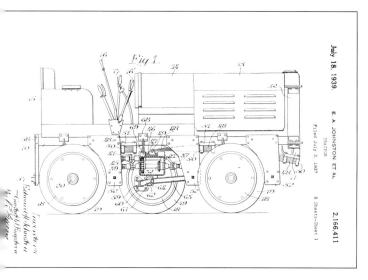

A patent showing the side view of the six-wheel tractor shows the gearbox mechanism for each wheel. Also apparent are the suspension springs and arms. *U.S. Patent*

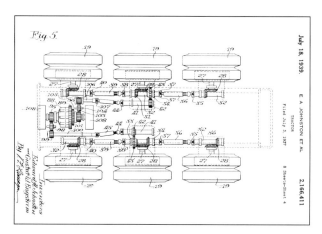

A drawing of the drive system of the six-wheel tractor. The power went into a splitter box (part 88) and then through shafts to another gearbox located in the center wheel of each side of the tractor. From there, power went both forward and backward to the other wheels. One wonders if the need for at least 16 driveshafts and gazillions of U-joints per tractor was the reason for not pursuing manufacture. *U.S. Patent*

A photo of one of the last IH six-wheel-drive tractors. This one has styling similar to what IH was trying in the experimental F-22s of that time period. This particular tractor is fitted with pneumatic tires and the independent suspension system. Tractor Q-3665 weighed 7,525 pounds and was photographed December 1, 1937. *State Historical Society of Wisconsin*

suspension systems were probably the main reason for the demise of the six-wheelers. There must have been some promise in them as IH went through at least two designs for the tractor. The second design used a separate, independent suspension system for each wheel, indicating a complete change in philosophy from the early 8-16 six-wheelers.

F-22

After the F-21, Harvester's engineers rolled out a new Farmall tractor that was considerably different from anything IH had experimented with previously. The F-22, first photographed October 21, 1936, was a streamlined, styled tractor with a one-piece front frame, instead of a beam-and-bolster type of construction for the other Farmalls. The tractor was really the first that IH had produced that could be described as streamlined. From the first F-22, several different directions were taken. At least one F-22 was produced with the beam-and-bolster construction, while IH's industrial designers experimented with several different styles for the sheet metal.

The first of the F-22s, this attractive tractor also has what appears to be a frame consisting of a combination of steel channels and castings forward of the transmission case. Only a few photographs of this particular model exist. The front portion of the tractor, including the steering, bolsters, radiator and grille, and starting crank, were patented by IH. The Gas Power Engineering Department went back to the steel channel frame briefly after this tractor. Q-3240 was photographed October 21, 1936. *State Historical Society of Wisconsin*

One of the different versions of sheet metal designed for the F-22. IH was just starting to discover industrial design, and some of the efforts seem to indicate they really had to try in order to get the hang of it. This tractor is fairly attractive, but the radiator grille wasn't very attractive. It probably wouldn't have been easy to produce, either. This tractor is also fitted with either a hydraulic or mechanical lift under the driver's seat. One good bounce and it looks like the seat would have hit the lift hard. Tractor Q-3575 was photographed June 24, 1937. *State Historical Society of Wisconsin*

A sheet metal version of the F-22 with a very different frame. Instead of twin channels ending at the transmission case, this tractor has a cast one-piece frame, with steel channels inserted into it. Implement mounting holes were drilled into the channels, reducing the cost and stress on the cast frame, a radical departure for the Farmall. Tractor Q-3580 weighed 4,520 pounds and was photographed July 29, 1937. *State Historical Society of Wisconsin*

Harvester was experimenting with power lifts and attaching systems at this time. Lifting systems patented during the mid-1930s included mechanical and hydraulic lifts. Harvester also tried the lifts fitted at different points on the tractor, including under the operator's seat and under the gas tank. Some experimentation was also seen under the frame. IH was always looking for the best combination. Unfortunately, the best lift, three-point, was already patented by Harry Ferguson.

The F-15 tractor featured the same general construction as the F-22, although some differences in the frame can be seen. This tractor could be seen as an indication that IH was seriously contemplating replacing the F-14 and F-20 with the new tractors. A very attractive tractor, but probably expensive to produce as shown. Tractor Q-3708 weighed 2,251 pounds and was photographed January 24, 1938. *State Historical Society of Wisconsin*

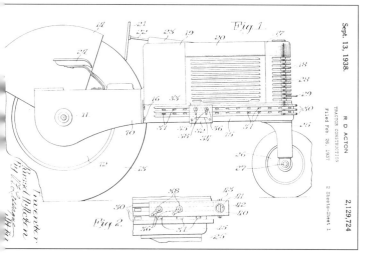

Russell Acton invented this interesting system of cultivator attachment, where grooves are cast into the forward frame (as opposed to the steel channel frame usual for Farmalls). A carrier is put into the grooves that can be slid into the desired position. A mounted implement then attaches to the carrier. It's a system with great flexibility, but would have probably been difficult to manufacture. *U.S. Patent*

The F-22 as it existed January 18, 1938. There has been some further refinement to the frame casting, the same sheet metal as the F-15 had, a steering wheel with a coating, and some other changes. This tractor was also equipped with steel wheels at one point for testing. Tractor Q-3682 weighed 3,820 pounds. *State Historical Society of Wisconsin*

Below
The 3-F tractor emerged in early August 1938 and apparently shifted IH's Farmall development from the solid-cast frame back to the channel and transmission case of the earlier experimental F-21 and F-22. The sheet metal was changed, with the front grille looking like the production Ms with the rear sheet metal transformed into something that could be considered plug ugly! Raymond Loewy was probably at work about the time of this photo. Tractor Q-3805 weighed 3,920 pounds and was photographed August 8, 1938. *State Historical Society of Wisconsin*

The W-42 tractor shows strong similarities to the F-22 and F-15 that were being experimented with during the early 1938 period, suggesting that IH was thinking about going into production with the new family of tractors. There are mentions of an O-22 at about the same time (no photographs have been found) so it would seem that there was a fairly complete family. However, the W-42 appeared only briefly as the entire family was changed. The 22 series was replaced by the 2-F and 3-F Farmalls with different frames. However, the basic design of the W-42 shows some similarities to the later W-9. Tractor Q-3781 weighed 5,918 pounds and was photographed June 21, 1938. *State Historical Society of Wisconsin*

The 2-F tractor. The 2-F saw the return to steel channels to form the front of the frame of the tractor. The redesign seems to have come from an August 1938 meeting at IH's Hinsdale Farm at which the entire tractor program was probably reviewed and some new directions mapped out. However, the appearance of the tractor certainly went backwards. Tractor Q-3827 weighed 2,690 pounds and was photographed October 26, 1938. *State Historical Society of Wisconsin*

The 3-F tractor. This tractor appears to be an exact copy of the 2-F, scaled up to a larger size. At a closer look, small differences can be seen, especially the magneto location, type, and drive. Tractor Q-3830 weighed 4,125 pounds and was photographed November 3, 1938. *State Historical Society of Wisconsin*

A second preproduction series 2-F tractor. Not too far from the H production tractor, the 2-F retained the different seat support and steering wheel supports, as well as other minor differences. The transmission case-front-frame meeting point also looks different in this view. Still, production would start soon. Tractor Q-3870 weighed 3,120 pounds and was photographed January 11, 1939. *State Historical Society of Wisconsin*

A second preproduction series 3-F tractor. The Loewy-designed sheet metal has been applied to this tractor, creating the look of the Farmall M. However, it isn't quite the production M, as can be seen by the seat support, steering wheel support and light bar, wheel equipment, and other parts. Yet, it won't be too long before production starts. Tractor Q-3847 weighed 4,290 pounds and was photographed March 2, 1939. *State Historical Society of Wisconsin*

Implement attaching systems were also experimented with. One of the early complaints about the Farmall system was that the implements were complicated and hard to attach. IH tried a variety of cast-in ridges and grooves in axles and frames to try to make the task of attaching cultivators and other carried implements easier. One system used ridges on the axle. A hook was snapped over the top of the axle, and another piece was bolted across the bottom to hold the hook in place. In another system, grooves were cast into the frame of the tractor. Hand cranks were arranged to slide in the grooves to provide an adjustable hooking system of great flexibility. Instead of the mysterious bosses and holes appearing on today's tractors, nearly any implement configuration could be handled by the groove system.

A New Family of Tractors

International started to move beyond experimenting with the F-22 by early 1938. The F-15 made its appearance by January 18. A smaller tractor, the F-15 weighed in at 2,251 pounds, while the F-22 weighed 3,820 pounds. A strong family resemblance existed between the two tractors, which used the cast frame. The wheel tractor version of the family made its appearance in June. The W-42 had a similar frame and sheet metal, notably the curious pinched-in cowling directly in front of the operator on both sides. The W-42 weighed in at 5,198 pounds. There is also mention of an O-22 (which had the same engine as the F-22), but no photographs of this tractor have been seen.

Letter Series Development

The Gas Power Engineering Department redesigned the F-15 and F-22. New frames more similar to the old 12 series system of beam-and-transmission case were used. Weight also increased, while the styling got considerably uglier. The names of the tractors changed as well. The F-15 became the 2-F, while the F-22 became the 3-F. Even this is a little strange, bringing to mind the question of what would have happened if Harvester had built a larger Farmall—it is doubtful that a tractor called the "4-F" would have sold well during World War II, or at any other time for that matter.

After a conference at the Hinsdale Farm in August 1938, where fresh requirements for the new tractors were laid out, the design team created the new F-series of tractors including the 2-F and 3-F. The new F-series tractors wore several versions of

The Farmall H with wide adjustable front axle, seen a few weeks before the start of production. The tractor now has the name and the look, yet there are a few parts that need revision, including the steering post and light bar, the seat support, and a few others. The extended starting crank necessary to clear the front axle in application is also pretty interesting—too big to fit in the toolbox while sticking out in front enough to potentially create a problem. The front axle weighed 425 pounds and was photographed May 31, 1939. *State Historical Society of Wisconsin*

streamlining, which gradually became closer to that of the production letter series tractors. Schemes were designed by IH's own industrial designers at first, but Raymond Loewy, a famous industrial designer at a time when industrial designers were household names, was brought in to do the final sheet metal plan. Loewy eliminated side curtains, producing an attractive, easy-to-service tractor that was also cheap to build. Loewy also redesigned the McCormick-Deering cream separators. He probably did the styling in the last half of 1938.

With the redesign, the tractors were nearly ready for production. Several preproduction series of each tractor were produced in the last half of 1938 and the first half of 1939. The names of the tractors changed, sometime between March 2 and May 21, 1939, becoming the now-familiar H and M. However, the big news was their little brother.

The Allis-Chalmers Patent Fight

Allis-Chalmers had brought out the very famous B, the first "baby" tractor, which kicked off a competitive frenzy among the other manufacturers. They looked on with jealousy as the B quickly racked up sales and accolades as the tractor that could finally begin to eliminate the horse on the average small farm of the day. The attractive streamlining of the tractor also had competitors struggling to keep up.

International's general office started hearing about Allis' baby tractor in March 1937. The letter came from the service manager at IH's St. Louis branch. He described two small Allis tractors that he had seen being tested in Belleville, Illinois. He said that the tractor developed about 9 horsepower at the drawbar and would sell for $500 to $600. Someone from International wrote back to sales manager W. E. Peyton suggesting he try to follow the tractor to determine the make of the motor and to get photographs.

Reports came in throughout that summer about the Allis B (or "bee" as it was referred to in one letter from Allis-Chalmers itself). Frank Bonnes and J. M. Strasser (Bonnes was coming from Chicago, while Strasser was from International's St. Louis branch) had a look at the two experimental Allis tractors in early April 1937. After describing the tractor, Bonnes gave his opinions about the Allis tractor and about IH's small tractor which up to now had been the F-12. Bonnes had done extensive investigations into the tractor needs of small vegetable farmers, of which there were immense numbers in those days (1,200 to 1,400 in the Indianapolis area, a similar number in the St. Louis territory). In Bonnes' opinion, the 12 tractors "just missed" this market and rising prices on the F-12 had meant that the small garden farmers would not make that investment. He also saw some problems with the Allis tractor, mainly with the life of the high-speed engine and the durability of the torque-tube main frame. But if the Allis B turned out to be durable, Bonnes thought that it "is the greatest threat to our F-12 business in the field today." In conclusion, he stated that IH should be designing a true one-plow small tractor and put the F-12 in the two-plow class where it belonged.

Another report about the same visit contained the quote, "The Allis-Chalmers is the most free of unnecessary parts of any machine either Mr. Bonnes or myself have ever seen." The report (dated May 6, 1937) from the two concluded, "Of the various tractors visited, we believe that this model may give us some competition." In a later letter dated July 15, 1937, Bonnes hoped that the engineering department was "losing no time in proceeding with the development of a small tractor to compete with this machine."

The Engineering Department indeed was wasting no time developing an answer for the Allis B. The main problem to be resolved was how to cultivate one row. Normally, a single row would pass directly

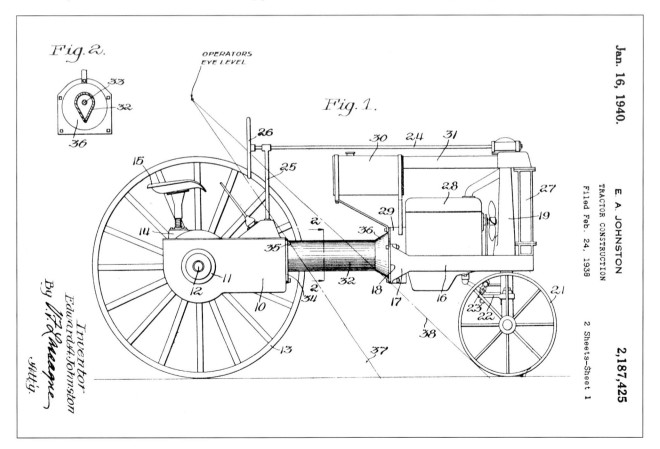

This tractor shows an IH attempt to patent a narrow torque tube to provide better vision for the operator of a cultivator tractor. This was probably done to try to get around Allis' tractor patents on the B and was ultimately unsuccessful. Still, it shows that IH engineers were thinking about a smaller, one-row tractor. *U.S. Patent*

under the tractor, and the operator's view would be blocked by the engine and transmission. Several different ideas to cure this problem were examined and patented. The first idea was using a narrow torque tube that would block much less of the operator's view. This idea had a problem. Allis-Chalmers pursued patent protection of the idea and got it, although IH was successful in getting a patent that probably offered little protection.

The Triangle Tractor

Another more original idea could be called "the triangle tractor." A patent, listing A. E. W. Johnson, a legendary IH implement designer, as inventor, shows and describes a tractor that is probably unique. The engine, transmission, a torque tube, and rear differential were laid out at an angle to the direction of travel, so that the entire assembly was moved out of the operator's way.

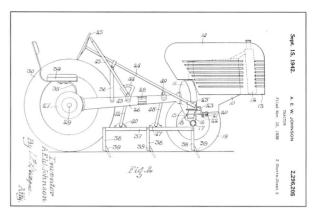

A side view of the triangle tractor. From this view, it still looks like something close to normal, although still unlike anything in the field. The cultivator system doesn't look too reliable though. *U.S. Patent*

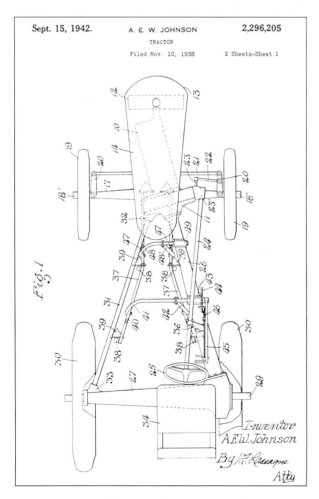

The top view reveals the triangle tractor in all of its glory. The engine, torque tube, and rear differential are all constructed at an angle to the direction of travel. While the drawings show the cultivator list lever and the throttle lever, they don't show the location of such thrills as clutch or brake pedals. It would have been interesting to see how they positioned them. This tractor shows great creativity and was probably a major step toward Cultivision. *U.S. Patent*

The other side of the tractor had a brace running from the front axle to the other end of the rear axle, forming the other long leg of the triangle. The rear axle had the differential on one leg and the brace on the other, forming the small side of the triangle. The operator was not on the centerline of the tractor, but instead moved off to one side, probably so that the clutch and brake pedals could be mounted on the torque tube. This idea would also show up on the other idea IH had for the one-row cultivating tractor: Cultivision.

Cultivision Development

Cultivision tractors had an offset driver's position on the right side of the tractor over an extended axle and casing. The entire engine/transmission/differential were slid to one side of the tractor, leaving nothing but the right front wheel in front of the driver. Having the most of the tractor on the left side, very near to the left wheels, led to some balance problems. Engineers eventually heavily weighted the right wheel, while the left was just a steel hub. Cultivision offered the driver an unparalleled view of the crop being cultivated, literally right before the driver's very eyes, instead of on each side of a central body. Cultivision was a novel idea, supplying an identifiable and salable design feature that would remain in production until 1979.

Models 1-F and F-10

Cultivision became the preferred choice of IH engineers, with two series of prototypes being constructed. The 1-F and F-10 differed mainly in the type of steering that they used, the 1-F having a rod running directly from the steering wheel to the front axle-mounted gear box, while the F-10 had the steering gearbox located under the tractor. After several prototypes had been built, the first series of preproduction 1-Fs (there is no record of preproduction F-10s, meaning those produced were

The F-10 tractor. IH's engineers produced two prototypes of a one-row Cultivision-type tractor. The F-10 differed from the I-F most obviously in the steering gear. The F-10 weighed 1,650 pounds and is seen April 21, 1938. *State Historical Society of Wisconsin*

The I-F had the steering wheel directly connected to the gearbox on the front axle (hidden in this view). Other details can be seen, such as a squared-off fuel tank, a different engine and torque tube construction from the later A, and a different throttle location. Tractor Q-3755 weighed 1,725 pounds and was photographed June 2, 1938. *State Historical Society of Wisconsin*

The "IF Tractor Wide-Tread Attachment" is the first sign of Farmall B development. The B was basically an A with an extended left rear axle and housing, steering gear, and front axle, and was produced on the same production line in the same group of serial numbers. The purpose of the B was to cultivate two rows instead of the A's one, hence the narrow front. While IH could make the B cheaper that way than by producing a different tractor, the B was much less flexible in tread settings than the F-12/F-14, necessitating the BN (narrow tread) and finally the Farmall C. Tractor Q-3807 would itself be converted back to a I-F (a prototype). Photo taken November 4, 1938. *State Historical Society of Wisconsin*

This is a second preproduction series 1-F tractor. It's starting to look like the production A, with a different steering wheel, steering wheel support, seat, a grille that doesn't look quite right, and other differences. Tractor Q-3807 (remember that number?) weighed 1,450 pounds and was photographed January 10, 1939. *State Historical Society of Wisconsin*

The WD-9 prototype on full steel. The WD-9 was the replacement for the WD-40, America's first production wheel diesel tractor. The frame shows similarities to the W-42 of a year and a half earlier. Tractor Q-4409 weighed 4,780 pounds and was photographed January 9, 1940. *State Historical Society of Wisconsin*

Right
The O-6 development ran into some snags. The first O-6s seen had a smooth exterior sheet metal. The next O-6 was seen with cooling holes and louvers. This tractor saw the holes carried out even further, indicating that the cooling problems were fairly serious. Tractor Q-4098 weighed 4,950 pounds and was photographed June 6, 1940. *State Historical Society of Wisconsin*

prototypes) was authorized August 25, 1938.

The small tractor program, which eventually combined features from the 1-F and F-10, had a serious problem shared to a certain extent with the larger tractors: Allis-Chalmers was there first. In planning the production of the A, International executives, mainly patent attorney Paul Pippel, had asked Allis exactly what Allis was going to claim in patent protection. IH obviously wanted to avoid violating Allis' patents or having to buy a license to use Allis' technology. However, Allis' patent attorneys at first did not consider the IH tractor to be a violation. There are several reasons why they may have said this at the time. They may have considered Cultivision to be enough of a difference from the basic Allis design to avoid interference. They also may have been unsure about which of the Allis patents would be granted by the Patent Office. For whatever reason, the Allis-Chalmers patent attorney, DeWein, told IH not to worry.

International proceeded with the design of the tractor, while keeping in contact with Allis. Allis specifically did not share with International the nature of the patent applications Allis had made. IH's patent attorneys searched the "prior art," that is, patents already granted. In this area, the attorneys concluded that IH was in the clear with the A's

The preproduction W-4. A tractor that had to wait until the Farmall line was placed into production, this tractor shared an engine with the H, although the frame had more in common with the F-15. Tractor Q-4104 weighed 3,240 pounds and was photographed January 9, 1940. *State Historical Society of Wisconsin*

This is a preproduction W-9 tractor. The design of the W-9 shows heavy influence from the W-42. W-series development probably waited until after the Farmall lines were in production. The W-9 and I-9 would see heavy production during World War II for government contracts. Numerous variants to meet a specific government contract would be developed—and may have already been under way for British and Canadian orders. Tractor Q-4196 weighed 5,420 pounds and was photographed August 2, 1940. *State Historical Society of Wisconsin*

The "Farmall A Tractor With High Clearance Attachment." This first AV had to overcome several hurdles due to the unique front axle design. The problems were overcome and the AV is a desirable tractor among collectors these days. Tractor Q- 4106 weighed 2,145 pounds and was photographed January 26, 1940. *State Historical Society of Wisconsin*

The "Farmall H Tractor With Cane Tractor Attach." is seen October 4, 1940. The H was the first letter series Farmall to receive the cane tractor treatment. Up to this point, F-20s were still being produced when a cane tractor was needed. Q-4087 didn't have a weight recorded. *State Historical Society of Wisconsin*

The "Farmall M Tractor With Cane Attachment" is seen here on December 10, 1940. It took a while for the gas power engineers to design the smaller production variants. The Gas Power Engineering Department was responsible for tractors, crawlers, power units, the small hopper-cooled engines, the whole refrigeration program, dairy products, the household equipment program, truck engines, cotton pickers, self-propelled combines, and various other products that involved engines or precision, high-tech engineering. All of the above just happened to be getting engineered or re-engineered at the same time! Tractor Q-4233 weighed 5,510 pounds and was photographed December 10, 1940. F-30 cane tractors were being produced at least until this date and probably for quite some time thereafter. *State Historical Society of Wisconsin*

design. However, the potential Allis patents still could throw the program for a loop.

DeWein soon changed his mind about whether IH's small tractor could be a violation, apparently at the urging of Harry Merrit, the man responsible for Allis' tractor program. Allis was then granted patents covering the basic design as several features of the Allis B. International had a look at the patents for the first time and didn't much like what they saw. Still, the IH A with Cultivision looked very different in some respects than the Allis B, which would get IH around Allis' design patent. Allis-Chalmers management did not buy this theory and told IH that they wanted several dollars a tractor from IH in license fees. IH didn't really want to pay that much in license fees and stalled the negotiations while a number of strategies were tried.

One strategy explored was to change the design of the IH tractors. (The entire Farmall line violated at least some part of the several patents Allis received.) Variations of the Farmall A design were tried, especially with the design of the clutch housing (which made it into production after being tried on an experimental tractor) and the gas tank. This was the time period shortly before the Farmall B was to be placed in production. IH's executives considered an entirely different design for the B that would avoid any infringements, although no pictures of an alternative B can be found. This may, however, be the start of the design of the Farmall C.

Another strategy was to disprove the validity of the Allis patents. According to law, a patent must be applied for within two years of the first sale of the innovation or product to be patented. IH did some simple math and realized that Allis-Chalmers may have cut the two-year time period a little too close. A patent research firm from New York was employed, and soon a private eye was searching for the owners of the first 100 Allis Bs to determine the precise date of sale. The private eye also started a friendship with a certain Allis-Chalmers employee with access to the build cards. The cover story for the investigation was that a fellow was interested in buying several Allis Bs, but wanted to talk to owners with the oldest (most used) Bs to determine how well the tractors held up. The investigation revealed that the date was indeed very close, but determining the actual time that Allis started to sell the tractor depended on definitions. However, the owners of the Allis Bs, while telling the age of their tractors, also told the investigator that those Allis Bs were one heck of a tractor.

Eventually, the experimentation with new, non-infringing Farmalls was halted as being "defeatist." Allis was getting tired of waiting to make an agreement and made a very nasty threat—if IH didn't come across with some money, Allis would go out and collect the money directly from farmers! The IH attorneys didn't take that threat seriously. However, the attorneys did threaten Allis right back with the revelation that Allis may have invali-

The "Farmall M Tractor (Diesel)" was probably the obvious result of the WD-6 engine being mated with the M frame, which shared the same gas engine as the W-6 and, therefore, was an easy installation. Although IH had experimented with diesel-powered Farmalls before, the MD would be the first to see production. Tractor Q-4220 weighed 4,760 pounds and was photographed October 18, 1940. *State Historical Society of Wisconsin*

dated its own patent by selling the Allis B too early. At first the Allis patent attorney laughed. But within a few days, the companies settled the fight. International Harvester paid Allis-Chalmers $35,000 for the right to use the Allis patents—a fraction of what Allis originally asked.

W Series Tractor Development

After the W-42, the conventional wheel tractors were virtually ignored as Farmall development, especially that of the smaller tractor, was the main priority. Few if any W prototypes were produced during the main push on the Farmalls, although the Gas Power Engineering Department probably kept them current to the Farmall efforts on paper. IH's resources to build experimental tractors were probably tied up with the Farmall program, pushing the lower sales conventional tractors back. After the letter series got into production in mid-1939, W series prototypes appeared quickly.

Engines and other parts were shared with the Farmall line in the W-4 and W-6, but caution should be used in referring to them as "conventional versions of the Farmall." The frames of the W series appear to be very similar to the frame of the W-42 and the F-22.

Filling Out the Farmall Line

The F-series Farmalls had evolved into a line including high-clearance and cane tractors. The letter series Farmalls did not have these variations when introduced, so a small trickle of the specialty F-30s and F-20s continued to emerge from the factory. IH's gas power engineers had quite a bit of work to do, so the cane tractors had to wait a little while. Nineteen forty saw the engineering done on the A, H, and M, eventually producing the AV, HV, and MV tractors. The high-clearance models reverted to using final drives, which lowered the axles and enabled normal tires to be used, while also lowering the center of gravity.

Diesel Development

International had experimented with diesel-powered Farmalls for years, but never got into production during the F-series. Yet progress was inevitable, and the M received a diesel engine in late 1940. This was probably a result of the production of the WD-6 with the same engine. Since fuel consumption of the larger wheel tractors was much heavier, and the larger tractors were much more tolerant of extra weight and expense, the larger tractors were equipped with diesel engines before the smaller, more price-reliant tractors.

World War II Development

Tractor development in World War II was limited by devotion to military projects, material supply problems, and the fact that IH had just put out a new line of tractors. Industrial tractors designed for military special needs were prominent, and so were conversions of the regular crawler line.

Farm tractor production was drastically reduced by material shortages. Pneumatic tires were definitely unavailable. Steel wheels made a comeback, which was no problem for the H and M, which had been tested and sold with steel wheels from the beginning. However, IH had never tested the A and B with steel wheels, intending them to always be rubber-tired tractors. When tires became unavailable, IH tested five As and five Bs with steel wheels after locking out high gear in the transmission. The new steel wheels were authorized May 15, 1942 (very shortly after the Japanese had captured most of the world's supply of natural rubber), and all tractors were shipped by May 27, 1942. The steel wheels were produced by the Gas Power Engineering Department.

By 1943–1944, International Harvester executives were starting to think about what the company would sell after the war. However, one very interesting development was the 20-miles- per- hour tractor. This tractor was clearly designed for industrial work. The demise of the I-14 created a gap that was not filled by the I-4. The I-4 was clearly a larger tractor that would not fit into many of the tight spaces and weak floors found in America's older factories, some of which dated to the mid-1800s. Even IH's vast McCormick Works suffered occasional floor collapses due to the increased weight of manufacturing machinery, while hall spaces in some cases were extremely narrow.

The answer that the Gas Power Engineering Department tried was a tractor apparently based on the Farmall A, but built much lower and with bumpers and attractive hubcaps. Although nothing came of this particular attempt, IH still needed a small industrial tractor to fill the gap left by the I-14 and to meet the market competition created by the Allis-Chalmers IB.

One of the wartime experiments was the "International 20 M.P.H. Transport Speed Tractor With Hiway Mower," which pretty much tells the story. A bumper, hubcaps, and fenders create a tractor that would probably be an extremely collectible tractor today if it had been produced. This tractor had a hydraulic system common with the later Framealls, with the large control valve system located under the steering wheel in this view. Tractor Q-4331 was photographed May 12, 1944. *State Historical Society of Wisconsin*

Chapter 9

After the War and Beyond

World War II drastically affected IH's agricultural tractor development. Production of most models was slowed, while the A/B was actually discontinued for a time. Resources were scarce, and the demands of IH's military production took a lot of engineering time. Still, the march of technology was inevitable.

Cotton Pickers

The cotton picker had been experimented with by IH for decades. Finally in the early 1940s, a workable design was produced, just in time to help relieve labor shortages caused by the war. IH needed a power unit for the new cotton pickers. In order to propel the picker, IH designed special versions of the H and M tractors. The tractors were fitted with reversed transmissions, reversed seats, and a modified cane tractor rear end. These tractors could be converted back into row-crop tractors with a special conversion package.

A further derivative of the cotton picker tractor was a tractor meant specifically for mounting combines. While corn pickers had been mounted on tractors for quite some time, combines (or in IH terms, "harvester-threshers") had been tried a few times (notably by Baldwin) but had not really caught on well. In the late 1930s and early 1940s, the farm equipment industry was caught up in an effort to design and produce combines as small and as cheaply as possible. This war was of course started by Allis-Chalmers and the "All-Crop" combines, but the other firms had been forced to follow. IH tried to build a combine that had no running gear and no power unit. Instead, it was mounted on a tractor with the engine on one side and long axles running under the combine to support it. The tractor was reversed, with the front axle (normally rear) being hinged so that the combine could be pushed into position. The base tractor was either an H or M.

Thinking about the Postwar

In 1944, L. B. Sperry gave a speech to the Society of Automotive Engineers (SAE). In this speech, which was widely reproduced, Sperry gave a synopsis of the current IH plans for postwar tractors. The purpose of this speech may have been to impress the War Production Board, which was responsible for allocating materials to International Harvester, in order to get the company to devote time and materials to engineering new tractors for after the war. Sperry gave five main requirements

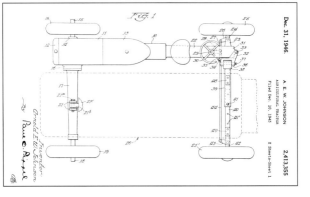

The combine-mounting tractor featured extended axles and a reversed position for the operator. Based on the International Harvester tractors modified for cotton picker mounting, this tractor was probably the start to extensive IH experimentals in a universal system of implements. A mounting pad for the combine is located on the rear axle. *U.S. Patent*

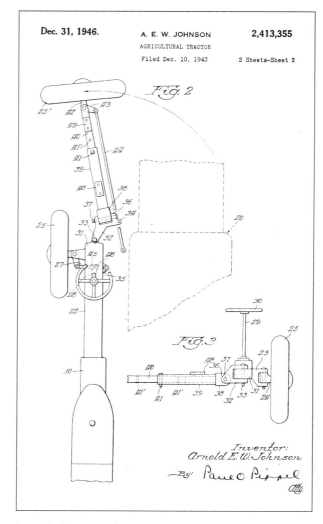

Instead of lowering the combine onto the tractor, or backing into it, the combine tractor had a hinged front axle. The combine probably was backed onto the tractor. The rear axle was still the driving axle on this tractor, but an engine-over-axle unit was probably used for the engine and transmission package. *U.S. Patent*

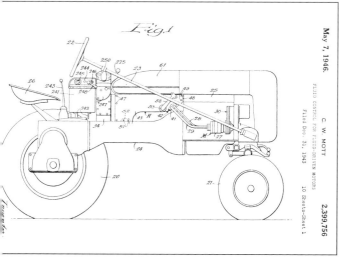

for a tractor to meet the postwar market:

1. Quick attach implements
2. Better controls for these implements
3. Operator comfort
4. Small tractors for small farms
5. Engine designs optimized for high efficiency with low-grade fuels

After his plea for more materials, Sperry gave some very interesting potential specifications, some of which fall into the gee-whiz category, even today. The engine would be "super elastic powered" with an oxygen-separating device providing pure oxygen without a need to have an air filter. Ignition for this engine would be from an indestructible catalyst.

Engine starting would be through radio-activation, with radio-activated headlights. This would eliminate storage batteries and electrical systems, which IH engineers traditionally tried to eliminate during the twenties and thirties with no-battery lighting systems and hand-crank starters.

Torque converters would be used, with automatic control of the vanes and stator blades. Automatic temperature controls would be present, as well as a manual control to provide reverse to avoid the need for reverse gearing and brakes. Infinitely variable speeds would be possible with a throttle control of the converter pump assembly.

Framealls

Before and during World War II, Ford-Ferguson's 9Ns and 2Ns had started to sweep the country by storm. The Ferguson system tractor was small, affordable, and the invention of the three-point hitch by Harry Ferguson had ensured the tractor power beyond its horsepower rating. The beauty of the three-point hitch was the geometry that transferred the draft of the implement into downward force on the rear wheels that increased traction, instead of causing the front end to rise. If the pull of the implement threatened to stall the tractor, the Ferguson System of hydraulics automatically lifted the implement slightly to avoid the stall.

Henry Ford's involvement ensured that the tractor would be produced and marketed at a very low price. With the advent of wartime rationing, Ford proposed to the U.S. government that the Ford tractor was sufficient for all needs, far more efficient in power and in

Left
A patent drawing of the early IH small tractor hydraulic package shows the huge size of the installation. The belt-driven hydraulic pump (27) is obvious, but the ram assembly (under the gas tank) and the control assembly (to the rear of the gas tank, under the steering wheel) take up a huge amount of space and weight. Later developments would reduce the size of the control assembly. *U.S. Patent*

many of which were already in inventory. In a less honest part of the memo, IH alleged that the only reason that the Ford-Ferguson needed the three-point hitch and the Ferguson hydraulics was that the tractor was not designed properly, and a properly designed tractor would never need a three-point hitch, an attitude that haunted IH until the 1950s. The government never did buy Ford's story, and IH tractors remained in production during the war (except the A and B, which went out of production for over a year), although numbers were greatly reduced and many low sales attachments were canceled.

It was obvious that Ford's combination of a hydraulic system in a small tractor was a great success. IH's manifold exhaust pressure lift system in the A and B worked, but required large amounts of maintenance

A rare photograph of the Frameall in action, this Frameall B was either owned or operated by J. S. Gibbs of Gordonville, Virginia, sometime in 1944. This tractor has the crossbar slanted in the opposite way from the patent drawings. It is unknown which series of Frameall this is, but it is probably one of the first series tractors. *State Historical Society of Wisconsin*

materials than the competition. He proposed that his tractor be the only tractor in the United States produced during the war, replacing all others in production and leaving Ford, of course, with a monopoly that would be hard to break in peacetime. Not only did Ford propose to replace all other tractors in production, he also proposed that all older tractors in existence in the United States be scrapped and replaced with Fords! The dangers of the early 1920s Fordson invasion, which IH just barely fought off, were back.

Harvester fought back with a memo pointing out that many tractors were more useful in certain situations, such as the large-scale farming in the West, for which the small Ford-Ferguson was hardly large enough. The memo also pointed out that many older tractors had plenty of life in them and it would be much more advantageous to supply these tractors with spare parts,

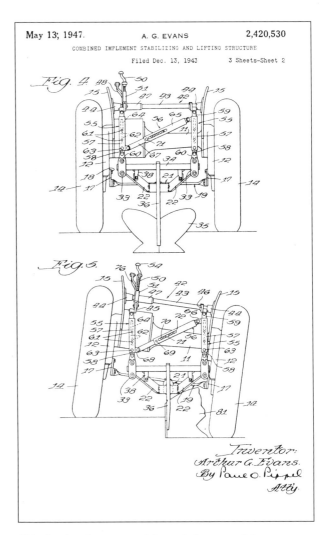

This drawing shows one of the main features of the Frameall—the ability of the hitch to have different characteristics for different implements. The middlebuster above is rigidly held straight up and down. The plow at the bottom is held looser, allowing the plow to take a different vertical angle than the tractor. The crank to make the adjustment is on the upper left side of the hitch (part 50). *U.S. Patent*

and often was ignored. The company read the writing on the wall and began experimenting with small tractor hydraulics in 1942 or 1943. The first experiments had a large bulky hydraulic system running off the transmission. The appearance of this system looks heavy, complicated, and expensive. Still, the system had enough promise to continue, and acquired the name "Frameall," an obvious take-off of the popular "Farmall." IH quickly became serious about producing these tractors, starting a preproduction series having a transmission-driven hydraulic system. The first series was authorized February 23, 1944, and 20 tractors were built by May 10. The first series differed from regular production As and Bs (the Frameall tractors were converted from regular production tractors taken from the assembly line) by having a new quick-attachable drawbar, new rear axle housings, new rear fenders, and a new steering knuckle post (A and AV only) to accommodate the Frameall system itself, which was a new hitch system meant to compete with the Ferguson System three-point hitch. The parts changed to accommodate the hydraulic system itself were a new clutch housing, new fuel tank support, new transmission case cover, and a new gearshift lever. Tractors could be fitted with the hydraulic system (treated as an attachment). When fitted with the hydraulic system, electric starting and lighting, radiator shutter and heat control, and distillate- and kerosene-burning attachments were different from the normal As and Bs.

The second preproduction series had an engine-mounted hydraulic pump, although the author has not seen a photograph of this tractor. The engine-mounted system was designed to reduce manufacturing costs from the earlier transmission-mounted system. The hydraulic control system was also redesigned to reduce costs. The new series was authorized July 7, 1944. Various setbacks prevented the tractors from being completed until March 22, 1945. The authorization for regular production of the Frameall series of tractors started to go through the authorization process October 28, 1944, before the second series preproduction tractors were completed (although a few may have been finished by then). The International A, meanwhile, was also redesigned to take into account the Frameall hydraulic changes. However, the move to produce the Frameall system was not approved by IH management, and development continued. An implement line was also being designed for the Framealls at this time.

The third series of preproduction Framealls had a dual ram hydraulic system, as opposed to the earlier series' single ram. New rear axles, new differential housings, and new platforms were designed to fit the new hydraulics. All second series tractors were modified into third series, and several more were built.

At one point, IH definitely planned to put the Frameall tractors into production. Specification lists

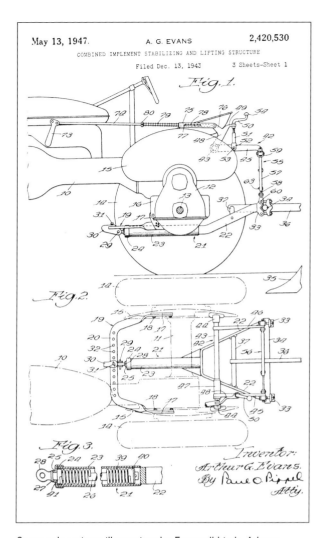

Some other views illustrating the Frameall hitch. A large spring-loaded device was part of the hitch and was located under the rear axle of the tractor (part 23, also illustrated in Figure 3 at the lower left corner). The arms at the upper part of the lift look similar to what was used in the Touch-Control system, but from there the similarity ends. *U.S. Patent*

were made, and for a while all changes regarding the regular A and B production were cross-referenced to the preproduction Framealls. There are reports of a Frameall C that was tested in Ohio. However, there also seems to have been at least a few tractors made with the Touch-Control hydraulics but without the hitch system of the Framealls. These tractors were tested probably in 1945–1946. It is still unknown, but what seems to have happened is that IH suddenly changed its production plans, probably in early 1947. Instead of the complete Frameall, the Super A with Touch-Control was produced. The suddenness of the change seems to indicate that there may have been a major mechanical or a patent problem. The Frameall implements were cut up and at least some of the tractors were returned to the factory, although IH provided parts support for other preproduction Framealls up into the early 1950s.

Industrial tractors also received some attention. This is the ID-501 tractor with a rock wagon. This equipment was probably based on the Heil system, which IH purchased from Heil at about this time. This particular piece of equipment was intended for the military. This photo was taken February 23, 1945. *State Historical Society of Wisconsin*

Cub Development

The early history of the Cub is elusive. The program was aimed at vegetable farmers in certain regions and was separate to a large degree from the regular tractor program. The aim of the program was a tractor that could replace two mules. According to one retired IH engineer, the Cub went through a redesign in 1942. The original Cubs used a welded chassis, which was changed to the more familiar cast iron of the regular tractors. Retired IH engineers describe the later design of the Cub as fairly easy. According to one, the Cub was basically a two-thirds scale of the Farmall A.

By 1945, the Cub was in nearly final form, and several preproduction tractors had been built. The tractor was introduced to the public at a dealer show during October 1945 at the Hinsdale experimental farm. The tractor was scheduled to be built at IH's new Wood River, Illinois, factory. The only problem with this was that the Wood River plant had not been built, and in 1945 materials for new factory construction were scarce to nonexistent. IH instead bought the ex–Curtis Wright plant in Louisville, Kentucky. The plant still required extensive expansion and conversion, so it was 1947 before the Cub could enter production. The delay allowed an extensive line of Cub implements to be prepared for sale by the time of the introduction.

Farmall C

The C seems to have shared many of the mysteries of the Cub. This tractor too was supposed to be built at the Wood River plant and had to wait for Louisville to be brought on-line. The starting date of C design is not known, but there is evidence that it may have started during the A patent fight, possibly in 1939. The documents state that at that time, a two-row tractor was being designed, and it was the perfect opportunity to change the design to one that would interfere less with the Allis-Chalmers patent than the A. Of course, the two-row tractor that was built was the B, which was an A with a different left rear axle and a narrow front, along with other changes. However, besides the patent problem, the B had other characteristics that were less than desirable with a two-row tractor.

The A/B design of the rear axle restricted how wide or narrow a farmer could set the tread. The brake housings and layout were the main problems. With the one row A, the ability to change treads was of less importance than with the B. A two-row tractor, the B had to have a large variety of treads in order to correctly fit the widely varied row sizes encountered, especially in the South. A narrow-tread version, the BN, was built with shorter axles, but required different mounted implements than the regular width B. There are photos of a BNW, which had a wide front end, with the narrow rear of the BN. On the C, IH engineers moved the brakes to the main chassis, enabling a large range of adjustment for tread. The C had hydraulics, as opposed to the standard production B. Bs had been built with hydraulics (the Framealls), but IH decision makers discontinued the B and went with C production.

Model IC

The discontinuance of the I-14 left a hole in IH's industrial tractor line, as the A and B were larger tractors. To fill the hole, IH Industrial Power Division employees proposed IH build an industrial version of

Another early Cub. This is probably a preproduction Cub issued to a real farmer to test the tractor. This one doesn't have the IH emblem on the top of the radiator. The Cub was ready for production in 1945, but IH didn't have a factory ready for it until 1947 due to postwar material restrictions. A factory was to be built in Wood River, Illinois, for the tractor, but a war-surplus factory in Louisville was already built and available for sale—so that's were it went. *State Historical Society of Wisconsin*

This is one of the Cubs shipped to IH's implement factories to aid in the development of implements for the Cub. It is unknown when this Cub was built, but indications are that it would have been 1944 or 1945, as the photo is taken in late 1945. This tractor was shipped to Canton Works, which oversaw much of the implement production for the Cub. The pale color is probably a result of IH's difficulty in obtaining red pigment for its paint late in the war. To stretch, IH added 50 percent yellow pigment, generating the pale color. *State Historical Society of Wisconsin*

the C. The IC was proposed December 29, 1947. In a memo written by E. A. Braker, he stated the IC "would provide versatile and economical power for railroads; industrial plants; highway mowing and maintenance; golf courses; park and airport care; street sweeping; side walk snow removal; and various material handling operations." He estimated that at least 2,000 a year could be sold and recommended a prototype be built. At least one prototype was constructed.

The proposed IC was to have the same engine as the C. Weight was to be 2,400 pounds and was to have a rear tread varying between 44 and 60 inches wide, with a wide front of 44 inches. The wheelbase was shortened to 65 inches from the C's 81.5 inches. Turning radius was increased by a foot over the regular C, but overall width, height, and length were shorter than the C. The tractor's estimated price was $1,075, as opposed to the Allis-Chalmers IB, which sold for $992.

Model ID-6A

During the early 1950s, IH's industrial power division planned to build the ID-6A. Not much is known of this tractor, other than it had a C-248 gas engine. Production of the engine was released, but was canceled December 2, 1952. No tractors or tractor engines were ever actually produced.

Intermediate M

The Intermediate M first appeared during the 1945 Hinsdale introduction of postwar tractors. The Intermediate M was apparently a preproduction program that was authorized for manufacture. Not many details exist of the tractor, but it is known that the Intermediate M's major difference was Touch-Control instead of Lift-All, as applied to the regular H and M. The hydraulic pump was located on the magneto drive. The rear transmission case also appears to be different, with more mounting lugs cast into the casing. There is a photo of an H with what may be Touch-Control and a different rear transmission case and PTO. Not much is known of this tractor either.

Super A Development

The Framealls, mentioned above, were the leading contenders to replace the A/B series of tractors. However, there seems to have been a very quiet program in the background of As and Bs being produced with Touch-Control, but without the Frameall system. *Red Power* magazine covered the Super B, which was an experimental tractor that was not a Frameall.

This is a Cub with an electric lift. A single rocker arm actuated both the front and rear cultivator gangs. The sheet metal above the grille also has the word "Cub" on it with a double-striped line to either side. This is a 1946 photograph. *State Historical Society of Wisconsin*

After the unknown calamity hit the Frameall program, the alternate program was quickly brought up to speed. The Frameall program had gotten to the point where specification lists and production plans were being made. All of these plans were apparently either canceled or used as the basis for the new Super A.

More Diesels

The idea of providing diesel engines for the entire Farmall line was looked at as well. Foreign production of tractors was again being restarted by IH in the late 1940s in England and Germany. The high price of fuels in these nations (as well as economic problems caused by World War II and the need to rebuild) encouraged diesel engines. IH produced a series of "conversion" diesel engines—diesels converted from regular gas engines. A, H, and M engines were designed and tested, with the M engine making it into production in Great Britain when tractor production started there.

Super C, M, and H Development

International marketing surveys done in the late 1940s and early 1950s pointed out what the farmers wanted—more power. Several experimental projects started to look at what the next line of IH tractors would consist of. Increased power gas engines were easy. Engines identified as "Super M" engines were already being photographed as early as 1949. Experimental Super H and Super C engines were not very far behind.

A survey done in 1949 polled dealers about what horsepower they thought the current line should have. They thought the two-plow tractor should have 27 maximum belt horsepower (23 rated belt horsepower), the two- to three-plow tractor should have 35 maximum belt (29.25 rated belt), and the three- to four-plow tractor should have 45 maximum belt horsepower (38.25 rated). The Cub would have 9 3/4 maximum belt horsepower, while there would also be a utility version of the two-plow tractor. The W-9 would remain in production. The dealers said that the Super A should remain in production until the new small Farmall entered production but then apparently be discontinued. The dealers wanted a properly designed wide front axle for the three larger Farmalls. International dealers were consistent in their dislike for the current IH wide fronts, which they considered heavy, expensive, and poorly designed.

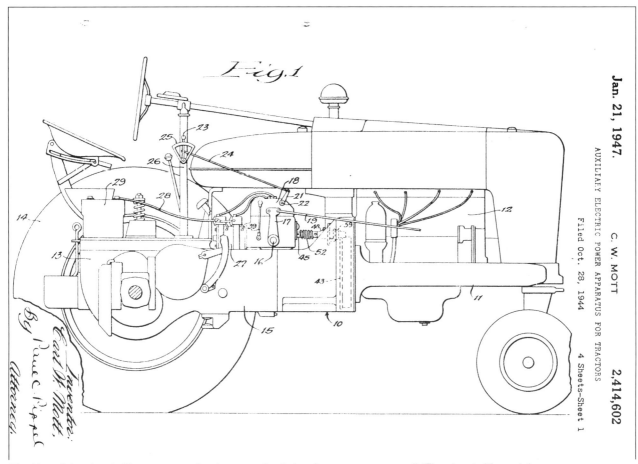

The idea of the electric lift was patented and apparently tried on larger tractors as well. The electric lift used the starter motor to provide the power for lifting. The lift was disconnected for starting purposes. The starter drive can be seen exiting the splitting and control apparatus (parts 45, 52, 48, and 49). *U.S. Patent*

The Farmall C's development is still a mystery except for the purpose: a two-row tractor with a greater range of tread adjustment than the Farmall B. This was achieved by eliminating the side housing and final drives, providing a straight axle that the rear wheel could slide in and out on. This is a preproduction C. *State Historical Society of Wisconsin*

Another survey was done in 1950. Corporate planners used the data to formulate yet another version of what the new tractors would be. The conclusions here are somewhat different from the earlier survey. IH dealers thought they needed a 41-horsepower M, a 31-horsepower H, a 24-horsepower C, a 19-horsepower Super A (if they needed one at all), and an 11-horsepower Cub. Many dealers recommended the discontinuance of the Super A. The addition of Touch-Control had made the tractor somewhat expensive for its small size. The one-row cultivators provided were another problem—many dealers recommended a two-row option. Others recommended more power to compete with the Ford 8N.

Similar complaints were made against the C—low power and high price, although price was less of an issue than it was with the Super A. One dealer pointed out that the Ford had 5 more belt horsepower but sold for $300 less in the Toledo area. The result was predictable; lost sales!

Perhaps the most interesting part of the 1950 survey was the section on the H and M tractors. Most of the dealers wanted a new transmission with more speeds in both tractors. The dealer in Sioux City reported, "We are badly in need of transmission changes rather than an increase in power. Larger motors would tend to increase fuel consumption on lighter loads, while a larger number of working speeds would give more economical use

A wide front preproduction C. The C, like the Cub, was announced to the public in September 1945, at a large dealer's presentation at IH's Hinsdale Farm. However, IH did not have factory capacity to build the tractor until 1948, at Louisville, Kentucky. *State Historical Society of Wisconsin*

of larger tractors on light loads.... The transmission speeds are our competitor's biggest talking point."

Dealers also wanted Touch-Control on the H and M. Some dealers wanted a larger Farmall altogether, with W-9- type power, to compete against Massey-Harris 44s and John Deere Gs. For all tractors, large or small, a constant running PTO was also wanted. More criticism was heaped on IH's wide front axle design and cost.

Reorganization and Expansion

International started to give the dealers (and customers) what they wanted. However, some changes in bureaucracy, leadership, and philosophy started to interfere with the activities of the tractor engineers. The change in bureaucracy was IH's reorganization in 1944.

Previously, there was the Gas Power Engineering Department, which handled truck, tractor, crawler, refrigerator, stationary engine, and other similar engineering. The Manufacturing Department handled *all* manufacturing, while the Sales Department handled all sales, and so on. The setup had worked well in earlier years, but by the mid-1940s was becoming a problem. The expansion of product lines, as well as the stress caused by World War II, started to cause management problems for IH. Executives simply had too much to worry about. Fowler McCormick started a program to separate IH into divisions according to product. Farm tractors became its own division, and so did Agricultural Implements, Trucks, Refrigeration, and Industrial Power.

This probably wouldn't have been as much of a

problem had the product lines not been expanding. Refrigeration needed money, as did the crawler and industrials. IH was trying to expand both of these lines. Engineering talent was divided, as well as money. The massive factory buying spree that IH went on after World War II was a result of the separation and soaked up money and delayed product introductions. Where before all tractors in North America had been produced in three plants—Milwaukee, Chicago Tractor Works, and Farmall—now Melrose Park and Louisville were added.

The whole mess meant that engineering money and priority for tractors started to fall behind. One IH engineer recalls that much of the money that IH placed into engineering went into the construction side, which was busy trying to produce a TD-24 that worked, as well as scrapers, large construction-type trucks, and other new products.

Early Advanced Transmissions

The search for better transmissions started at least in the mid-1930s. The idea of a live PTO was of course first tried out with the McCormick Automower in 1900. The idea never seems to have gone too far away. In 1930, experimental 10-20s were built with a variety of means of getting power from the front end of the engine back to the implement, but nothing was actually released for production. These experiments probably pointed out that a live PTO had to pass through the transmission.

The next step is a patent applied for in 1935. Ed Johnson, David Baker, and Clifford Rogers applied for a patent that combined a six-speed forward and reverse transmission (a shuttle-type transmission) with a continuous (independent) PTO. A belt pulley drive was also part of the assembly. A shaft was rigidly connected to the motor flywheel, which carried a sleeve shaft that

The "International Industrial C Tractor." IH documents reveal that even into 1948, IH's salesmen still felt the loss of the small, popular I-12/I-14 tractor. The industrial C was an effort to meet the market for a small industrial tractor that had only gotten tougher after the end of World War II. This is a Melrose Park photo. Melrose Park had taken over development of industrial lines after the re-organization of IH along product lines in 1944. This tractor was photographed June 30, 1949. *State Historical Society of Wisconsin*

The Intermediate M surfaced briefly in late 1945 at the Hinsdale, Illinois, dealer introduction of new models. The Cub, C, and Touch-Control A were introduced at this show as well. The engine-driven hydraulic pump can be seen, as well as a control layout that was very different from the regular M. Very little else is known about this tractor, and few other photographs are known to exist. *State Historical Society of Wisconsin*

received power from the flywheel via a clutch. The rigidly connected shaft passed through the transmission to the rear wall, where there was another clutch and gearing to the PTO outlet shaft. The sleeve shaft carried the transmission gears. Power was transmitted through the gears to another shaft, which was connected to the differential. The patent does not state whether the drive was by wheel or crawler track. The idea of using a fixed shaft for the PTO, with transmission gears on an outer sleeve shaft, is basically the method used when production of the Independent PTO began 20 years later.

Dual-Range Transmissions

The first IH tractor with a dual-range transmission was the TD-18. This tractor had a lever mounted on the dash that could be moved on the fly to select a new range of gears.

Another interesting patent was applied for in 1943 by Clifford Rogers, William Bechman, and Joseph Ziskal. In this patent, they devise a system to shift between gear ranges by the use of the clutch pedal. They state that in previous systems, it was necessary to clutch and to move two levers in order to switch transmission ranges. With the use of the new system, the operator could engage or disengage two clutches simultaneously, or each clutch selectively. A reduction gear set was located within the clutch housing. A sliding sleeve shaft was used to move gears into contact with the reduction gears. When not in operation, the sliding and main shaft were locked up for direct transmission of power. This transmission may have been meant for military vehicles.

Another attempt was made at a better transmission in the M-8 and H-8 tractors. Not much is known about this program, despite the existence of an M-8 tractor that

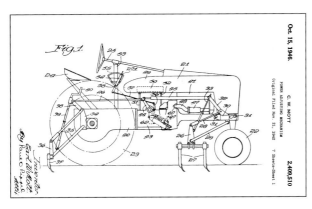

The planning for the Intermediate M had apparently been going on for quite some time. This patent, applied for in late 1942, shows the engine-driven hydraulic pump (part 47) as well as the rams, arms, and controls for the system. *U.S. Patent*

somehow survived the years. These tractors had multi-range transmissions with a total of eight forward gears. The tractors also had live hydraulics and a live PTO.

Torque Amplifier

The torque amplifier (TA) is perhaps the technology most attached to the IH name. IH introduced the TA with a splash in 1954, heralding it as the answer to the farmer's needs. The tractor lines were renamed to honor its introduction, while a massive advertising blast went out to convince farmers to try it out.

The torque amplifier system used a clutchable planetary gearset within an expanded clutch housing. The normal transmission was located behind the clutch housing. Normally, the clutch would be engaged, and the planetary gearset locked up, providing a direct drive to the transmission. However, when the operator desired a lower speed or hit a tough spot in the field,

A photo of what would later become the Super A. This tractor has the upper arms similar to the Frameall, but with the chain leading from the arms to the plow instead of solid linkages like the Frameall. The early control system (circled by an IH employee intent on making sure this photo wasn't used in advertising by mistake) identifies this as probably a 1943–1945 photograph. *State Historical Society of Wisconsin*

A 3x4 "convertible" diesel engine designed for the A tractor. Doing anything to the magneto on this tractor would have been a miserable process. IH made several attempts to build a 3x4 diesel engine over the years, but didn't get too far. This photo was taken January 2, 1946. *State Historical Society of Wisconsin*

the clutch was released, unlocking the planetary gearset. The gearset provided not only a reduction in speed, but also an increase in torque, the force that spins the wheels.

An early version of TA used a band type apparatus to control "wind-up" within the TA, which made it difficult to shift gears. However, as the M version of the torque amplifier neared construction in 1951, the band type apparatus had to be discarded. An overrunning clutch was then used, which required a revision in the lubrication of the system, which in turn required a complete redesign of the mechanism and clutch housing. The new lubrication system was such a difference that it received a different patent from the rest of the system. The system finally saw production in 1953.

The Independent Power Take-Off (IPTO) made a comeback in the early 1950s, prompted by the development of the system by other manufacturers. IH dusted off some of the old ideas, eventually mating a planetary gearset with the separate shaft running through the transmission. Although the first of the new patents was granted in 1941, it still took a few years to

The "Super M Convertible Diesel Engine" was apparently an attempt to update the engine used on the MD, but may have also been an effort to produce an engine more suitable for British production, which started at Doncaster, England, in 1949. Photo taken June 8, 1949. *State Historical Society of Wisconsin*

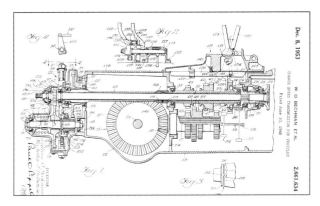

An unusual transmission dating from 1948 shows that IH was already thinking about multi-range transmissions and the Independent PTO (IPTO). In this transmission, power was transmitted from the engine to the PTO, and then back over a sleeve shaft to the transmission. A lower gearset (in phantom layout) can be seen at the right lower side of the transmission. Not much else is known about this transmission. *U.S. Patent*

put the system into production, even in the face of strong demand and competition.

The Great Transmission Chase

The mid-1950s saw an enormous amount of IH farm tractor engineer attention going into transmissions and draft control. The new advanced engineering activities within IH separated out the gee-whiz activities from both the engineering of production tractors and from the engineering support needed by tractors already being produced, or even those tractors that had stopped production. The most dramatic development went into transmissions. It should be noted that in the 1950s, the tractor engineers had less and less to do with actual engine development.

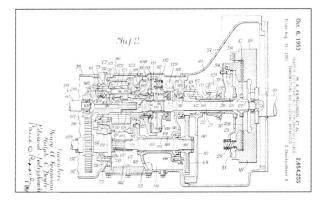

The torque amplifier used a planetary gearset and clutches to provide a low range of speeds that could be moved into on the fly from the higher speeds. The patent was applied for in 1951, but a series of last-minute changes prevented introduction until 1953, in the Super M-TA tractor. In this drawing, the TA is in the upper left side of the drawing. *U.S. Patent*

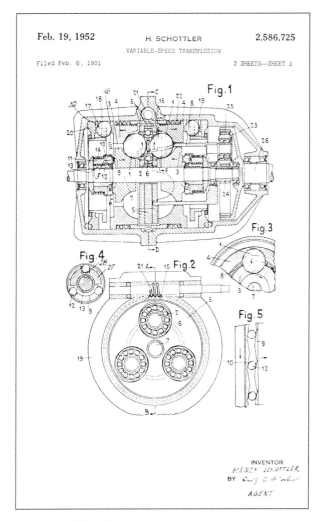

In the late 1950s, IH's advanced engineering organization embarked on a whole series of experiments with tractor transmissions, including planetary transmissions, Chevrolet Hydramatic automatic transmissions, torque converter transmissions, and so on. This design was probably one of the strangest. Henry Schottler worked for a major ball bearing producer examining many new uses for ball bearings. A variable speed transmission was one of them, and IH picked up the idea, providing Schottler with support, space, and assistants. The basic idea was to change the angle of rotation of the large ball bearings, changing the ratio of rotation of the parts in contact with the balls. This patent is believed to be the basis of the later experiments. *U.S. Patent*

One IH patent makes clear that the planetary gearset was being examined for even more use. One transmission patented was very similar to the later Ford Select-O-Speed. Three planetary gearsets combined to give the tractor several forward and reverse speeds.

Another patent showed that IH was examining the use of a torque converter, much like Allis-Chalmers in its large crawler tractors. The patent lists several disadvantages to the use of torque converters and the rather elaborate methods that IH used to overcome these disadvantages. The transmission ended up fairly complex, while the traditional low

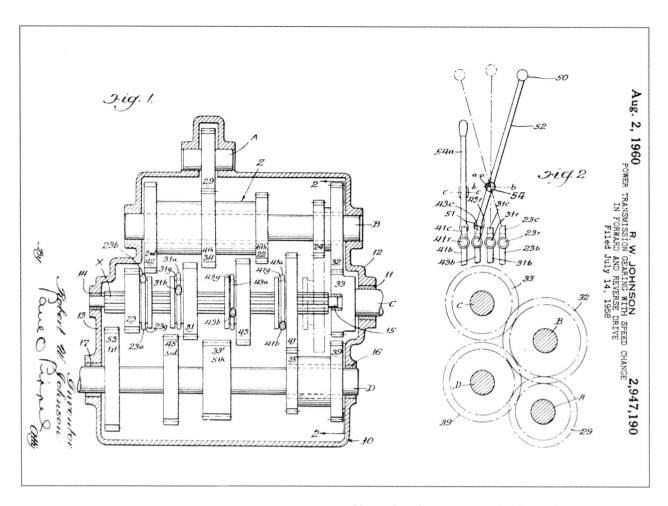

Yet another idea experimented with was this synchromesh transmission. Again, not terribly much is known about this transmission. *U.S. Patent*

The HT-340 originally started life as a regular 340 Utility tractor before an experimental hydrostatic transmission was added, producing the HD-340. Next, the conventional diesel engine was replaced by a turbine engine. The need for a bigger fuel tank, combined with a much smaller engine and elimination of the radiator, enabled a futuristic restyling to be done. The basic tractor is in front, with the stylists' drawings for the restyling in progress in back. *State Historical Society of Wisconsin*

efficiency of the torque converter while under high loads was insurmountable.

International started to examine the use of hydraulics in transmissions in the 1930s. At first, the main examinations were of using hydraulics for shifting. Soon, engineers were looking at an all-hydraulic transmission using pumps driven off the engine and hydraulic motors located near the crawler drives. The use of hydraulics would have eliminated at least some of the need for steering clutches and brakes. The early types were probably hydrokinetic, using the flow of low pressure hydraulic fluid to transmit power.

Another different type of hydraulic transmission was probably produced in small numbers in World War II. A few ID-9s produced for the Navy had torque converters similar to those used in car automatic transmissions. A photo of the installation exists, stating that the design was special for the Navy, although the caption does not state what purpose was being served. It is known that the Navy was especially interested in torque converters during World War II.

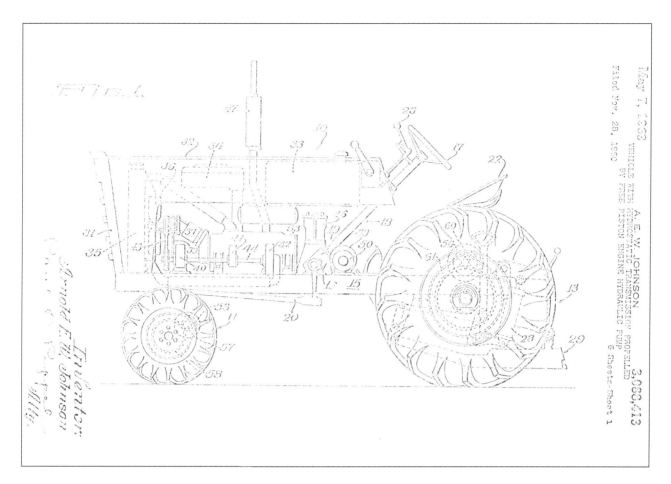

The free piston engine tractor used a free-reciprocating piston moving at very high speeds to generate exhaust gasses to rotate a turbine, which in turn powered the hydraulic pump for the hydrostatic drive. The engine and transmission package ended up being much smaller than the normal engine/transmission. The speed control is part 23, the fuel tank is 33, 37 is a hydraulic motor to power the cooling fan, and 49 is an air compressor powered by hydraulic motor 50. The compressed air was used to start the engine, 63. *U.S. Patent*

All-hydraulic transmissions reappear after World War II. International Harvester patented another all-hydraulic system of drive and steering in the early 1950s, but the patent states the main aim of the invention was at self-propelled hay swathers and harvester-threshers. Extensive work also went into construction equipment transmissions and power steering.

Another transmission design won the title of strange and novel. The Schottler transmissions used ball bearings in races to transmit power. The ramps/races moved the bearings in order to vary the ratio. Schottler came to IH to work on the design. Eventually, Schottler and the IH engineers assigned to the project went to a power split type transmission, where part of the power was transmitted through the ball bearing transmission, and another part transmitted mechanically. Although much experimentation was done, the project eventually died in the late 1950s.

The automatic transmission was another goal of IH engineering, whether through hydraulics or through mechanical means. One attempt involved the Hydramatic transmission, such as the one used in cars. A tractor was constructed around the transmission and tested extensively. However, a problem with the transmission's application to tractors couldn't be resolved. When shifting, the transmission sent a shock through the tractor and implements that was too great to deal with. New efforts were launched to create a new transmission that would replicate the IH torque amplifier transmission's characteristics, using two hydraulic clutches. This transmission also suffered from several problems. Another all-new transmission was designed to fit the same dimensions of the previous two transmissions, but this transmission worked. This multi-range powershift transmission was the 706/806 transmission. As the 706/806 came closer to production, however, there were several new problems to overcome. One demand was for more horsepower, so the clutch had to be expanded. The original tractor also used a planetary final drive. This also had problems, so a last minute redesign was made just before production began.

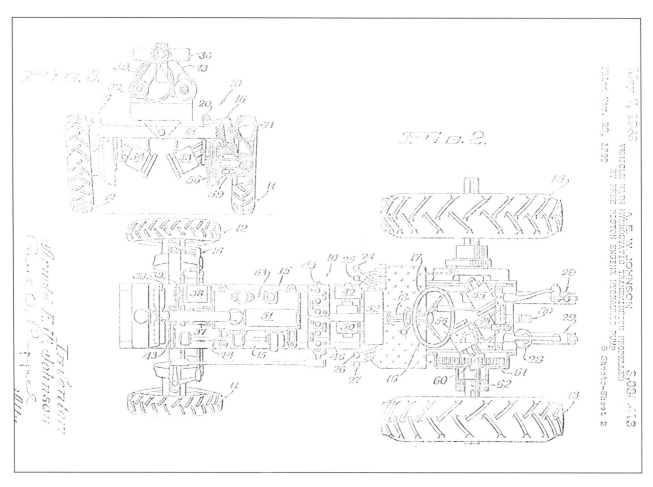

Front and top views of the free-piston tractor reveal that each wheel of the tractor had its own hydrostatic motor. The front wheels had motors 53 and 54, while the rear wheels had motors 55 and 56. The PTO was even driven by a hydraulic motor, part 48. The levers 24 through 27 controlled the lifts and PTO. *U.S. Patent*

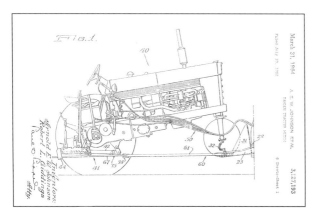

Of the two types of tandem tractors developed or examined by IH, this was the preferred form. The "pan type" of linking the tractors together required a minimum amount of modification to the tractors, allowing easy linkage for the times when the farmer wanted to combine tractors, such as for plowing. The front tractor hitched up the pan (part 23) in his Fast Hitch, drove the second tractor onto the pan, attached the chain (part 32) that kept the rear tractor from rolling or bouncing off, then raised the hitch and started attaching the hydraulic and hitch connections. *U.S. Patent*

Hydrostatics

Hydrostatic transmissions began to be investigated shortly after World War II. There was nothing really new about the hydrostatic idea. Harry Ferguson tried using a hydrostatic hydraulic lift when he was experimenting back in the teens and twenties but had difficulty controlling the depth of the implement—the hydrostatic mechanism was too sensitive and control was difficult. However, for transmission of power, hydrostatics were an easy way of varying the speed of a transmission and began to be used extensively in the 1940s, especially in construction equipment and in machine tools. (Among the early developers was Gisholt Machine of Madison, Wisconsin, famous for turret lathes and a corporate relative of Fuller & Johnson.) IH began looking into hydrostatics in the late 1940s, and an IH employee gave a presentation about hydrostatics' potential in tractors and construction equipment at the tractor meeting of the SAE in Milwaukee in 1954. During the presentation, he described the development of a hydrostatic steering system for a four-wheel-drive tractor, which was probably for the construction market.

International research engineers continued to work with hydrostatics throughout the 1950s, with applications eventually being made in several product lines, including power steering in tractors. Using the hydrostatic transmission in a tractor remained a difficult goal. In the late 1950s, the engineers took an International 340 off the assembly line, removed the transmission and made a few other drastic changes, and installed a hydrostatic transmission. The original tractor's gas engine was retained, and the resulting tractor was called the "HD-340." Tests were performed, and then the tractor underwent another drastic modification, receiving a gas turbine engine. International had experimented with its own turbine engines since at least the mid-1950s and later bought Solar, a company with an established record in the small turbine industry. The resulting tractor was called the "HT-340." A third set of modifications ensued when IH's industrial designers fitted the tractor with all-new sheet metal (and a much larger fuel tank!) to take advantage of the turbine's small size. Although the turbine/hydrostatic combination worked well (the higher rpm of the turbine allowed smaller parts to be used in the hydrostatic transmission), the turbine itself was loud, and fuel consumption was incredible. The HT-340 made a brief splash of publicity and was donated to the Smithsonian.

Contrary to public belief, this was not the end of the turbine tractor and the combination with hydrostatic transmission. Patent documents reveal that there was another program with another, equally advanced engine: the free piston. In a free piston engine, the piston reciprocates back and forth rapidly. The piston is not connected to a crankshaft. The main goal of the piston and cylinder assembly is to produce exhaust gasses. The exhaust is then passed through a turbine, which produces the actual power and is connected to the transmission. In the IH designs, the turbine of the free piston engine was directly connected to the pump end of a hydrostatic transmission. There are detailed drawings of this tractor in patent records, but no known photographs exist.

Hydrostatic experiments continued, both in France and in the United States. The French experiments looked at two motors, one on each wheel. A control system was devised to turn one motor at a higher speed than the other to assist steering. The U.S. developments looked at the design of a hydromechanical transmission. This transmission tried to overcome the efficiency problem of the pure hydrostatic transmission by transmitting part of the power mechanically. As speed increased, the power transmitted mechanically increased until top speed, at which the hydraulic transmission would be locked up and the mechanical linkage would transmit power at engine speed.

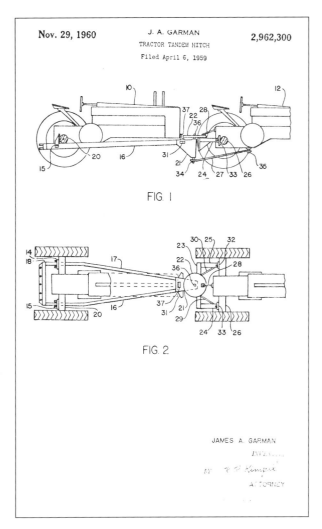

The earlier, less preferred version of tandem tractors involved removing the front wheels and axles of the rear tractor. Both versions of the tandem tractor were difficult to handle and were killed by the development of larger tractors and turbochargers. *U.S. Patent*

Chapter 10

Crawler Development

International Harvester's involvement in the crawler business is not as well documented as IH's tractor development. The first crawler development known of is that based on the IH 8-16 chassis as mentioned in chapter 3. At that time, IH could not produce a crawler mechanism that could take the stresses and wear that the rollers and guides were subject to. For a while, there seems to have been little crawler development at IH.

The introduction of the 10-20 and 15-30, especially the industrial versions, in the early twenties placed IH further into the industrial market than it had ever been in the past, just as the industrial market took off. In addition, the growing importance of California to tractor sales placed IH into another market where the ordinary wheel tractor could not perform as well as a crawler tractor.

Several firms started converting the basic 10-20 and 15-30 chassis into specialized industrial machines. Crawlers were no exception. IH started carrying the crawler conversions in special catalogs that listed the "allied" equipment that outside manufacturers made for use with the IH industrial power line. There was no doubt that the time for crawler tractors had arrived. IH's development of crawler tractors based on the new industrial tractors probably started around 1925–1926.

Development focused on both the 10-20 and 15-30. Several steering mechanisms were tried. The 10-20 received the unusual steering clutches and levers next to the operator. This went into production as the 10-20 TracTracTor. However, development of this system did not proceed much further.

The 15-30 started out with a similar system, but soon was converted to a tractor with the steering brakes placed in the chassis under the operator's seat at the rear of the tractor. Removable covers protected the mechanism while enabling easy service access to the brakes and clutches, a major selling feature in later years.

Although the 15-30 TracTracTor was not produced as such, development was soon affected by changes taking place in the conventional tractor line. The Increased Power program of boosting engine horsepower was built into the crawler prototypes along the same lines as the Ag tractors. The 15-30 first received an improved four-cylinder engine, which was soon replaced by the six-cylinder engine of the W-40. The 15-30 TracTracTor then became the T-40. Diesel engines were also being

IH struggled for years to get a crawler tractor into production, finally succeeding in 1928. The time to enter the market was right, as the Fordson had just left production. The many track conversions offered for Fordsons were left without any new tractors to convert for a few vital years. The IH 10-20 Track Layer, photographed on April 30, 1928, would fill that market. However, IH would later abandon this steering system. *State Historical Society of Wisconsin*

The 15-30 Track Layer experiments probably started about the same time as the 10-20 experiments, but never entered production as such. Instead, experiments continued for years as IH developed a different steering system and tractor layout. Here is an early 15-30 Track Layer, with the final drive exposed. This photograph is dated June 9, 1928. *State Historical Society of Wisconsin*

Here is another early 15-30 Track Layer showing that the steering system was not the only idea IH was exploring. The very wide treads were probably aimed at better flotation, or possibly increased stability. This photo was taken January 11, 1929. *State Historical Society of Wisconsin*

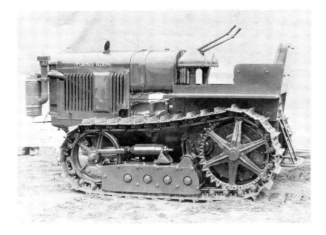

Here, the 15-30 is now named the TracTracTor and is showing some definite signs of evolution. The track roller assembly has been revised, as well as the air cleaner location. Many of the other changes seem to have much in common with the 15-30 increased power program, as well as the changes involved with the 22-36 modifications to the regular 15-30 wheel tractors. Tractor Q-883 weighed 10,000 pounds and was photographed December 16, 1929. This tractor still has the older steering clutch location, although the new revised layout was being experimented with by this time. *State Historical Society of Wisconsin*

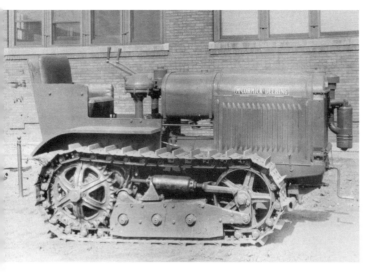

Left
The 10-20 TracTracTor also was being revised at the time. Again, the air cleaner position is revised, as well as the sheet metal and seat. Still, this tractor has the old steering clutches. This photo of tractor Q-856 was taken October 18, 1929. The tractor weighed 7,290 pounds. *State Historical Society of Wisconsin*

Here is one of the first photographs of an IH crawler with the revised steering clutch and brake location. This 15-30 TracTracTor has the older undercarriage, but the newer clutch layout. Tractor Q-834 weighed 9,910 pounds and was photographed September 3, 1929. *State Historical Society of Wisconsin*

Here is the Model 30 TracTracTor with what looks like an all-new undercarriage. The air cleaner has been moved to the side of the radiator, while the space under the fuel tank has been faired in with some sheet metal. Tractor Q-1275 weighed 10,040 pounds and was photographed August 27, 1930. *State Historical Society of Wisconsin*

tried out in the T-40, resulting in the TD-40, the tractor IH needed to compete with the diesel Caterpillars.

The 10-20 TracTracTor obviously was not a permanent solution to IH's crawler needs, so a replacement small crawler was developed, apparently as part of the small tractor program that resulted in the F-12 and 12 series conventional wheel tractors. The T-15 was built using the general layout of the T-40 with the clutches/brakes accessible from the rear of the tractor. The same chassis was used for an experimental wheel tractor that used the crawler system of steering. Twenty-five T-15s were produced (probably a preproduction run) before the tractor was revised and renamed the T-20. The small crawler, although apparently part of the small tractor program, never had the smaller 3x4 engine that the 12 series tractors used, probably due to the larger power requirements of the crawler. In fact, the power for the T-20 seems to have been something that IH was unhappy with. Both a six-cylinder gas engine and a four-cylinder 3

Tractor Q-1276 was a little better painted and decaled than its slightly older sister, Q-1275. Also appearing are the unique cooling louvers on the top of the tractor engine compartment. This tractor weighed 10,100 pounds and was photographed September 27, 1930. *State Historical Society of Wisconsin*

3/4x5 1/4 diesel engine were tried in T-20s in September 1933. These experimental tractors also had different track suspensions.

By the mid-1930s, the T-40 and T-20 were both in production, marking IH as a major crawler producer. Still, other tractors were needed to fill out the line in order to compete with the more established manufacturers such as Caterpillar, Allis-Chalmers, and Cletrac. The T-35 was the only tractor from the effort to provide more sizes to reach production. However, the need for a tractor larger than the T-40 prompted research into two larger sizes.

These tractors needed powerplants larger than anything IH had in production, so engine internals from Waukesha Motors were used in IH-produced blocks to provide the powerplant. New, larger transmissions and track assemblies were also needed. The new crawlers probably represented an enormous outlay in research money for tractors that were always going to sell in low numbers, and that did not share any parts with farm tractors that could help spread out research costs. However, the new engines could and were sold as power units, although the actual crawler tractors never were sold until a massive redesign.

The T-80 first appears in early 1935. This tractor was clearly huge, although not much more is known about it. A six-cylinder engine was used. The engine had two three-cylinder heads, a definite departure from IH practice. The operator's seat and controls appear to be offset to the right side of the tractor. Dual air cleaners were used. A look at the control levers indicates that a dual-range transmission may have been used, possibly with hydraulic controls.

The T-65 appeared about the same time as the T-80, and basically looked like a scale model of the larger tractor. However, the T-65 went through a partial redesign

This is the "redesigned" Model T-40 TracTracTor. Parts of the undercarriage are again revised. Visible on the rear of the tractor are the access covers for the brakes and clutches. This photograph was taken June 13, 1931. This is probably one of the last T-40 experimentals with a four-cylinder kerosene engine. *State Historical Society of Wisconsin*

shortly after. The next T-65 was a diesel and had a redesigned air intake system, as well as a new radiator guard and extensive changes to the track assembly.

By 1937, the crawlers went through another redesign in the large sizes, and some new designs for the smaller crawlers start to appear. The 60 became the 65, receiving what appears to be a complete redesign of the sheet metal and frame, and extensive revisions to the track assembly. The T-36 appears shortly after the T-65, looking again like a smaller version. Both had a curios control layout, with several of the engine control levers extending almost a foot back to the operator. Another version of the TD-60 appeared by mid-1937, looking almost identical to the earlier T-65. However, on close examination, the photos reveal several changes, including new sheet metal for the operator's seat, another redesign of parts of the crawler assembly. Another redesign of the TD-60 appears in December 1937 with a drastic change—streamlining.

The engine was now completely enclosed, and a revised grille was displayed. The effect is very similar to some of the F-22 experimentals being put out and may have been influenced by Raymond Loewy. The tracks also show signs of yet another revision. By January 1938, the TD-36 also received similar changes.

The TD-25 made its first appearance in June 1938. On this tractor, the engine side panels disappear, although the rest of the tractor is clearly based on the TD-36 and TD-60. Another TD-36 appears at roughly the same time,

with a wide-tread attachment, as well as accessories. IH was clearly readying a complete line of crawler tractors for production. In fact, the tractor did enter production, with the TD-25 becoming the TD-6, the TD-36 becoming the TD-9, and the TD-60 becoming the TD-14. The renaming occurred in late 1938.

Before and during World War II, the entire crawler line saw hundreds of changes, small production versions, and experimentals produced. The military was a primary customer, with myriad tractors produced for different duties in different areas. The problems of operating a tractor in the tropics or in water were far different than a tractor destined for the Arctic. Extra cooling was never necessary in the Aleutians, while a heated operator's cab and extra batteries for cold weather cranking were seldom needed on Guam. Other, more domestic needs resulted in a small number of tractors for agricultural use. The TD-6 orchard, for instance, was first experimented with in January 1941. The look and method were very similar to an earlier orchard version of the T-35. Some T-6s received an unusual modification for pea harvesters. On these tractors, the operator's seat and controls were rotated to the right 90 degrees. Presumably, the operator had to look over his left or right shoulder continuously to operate the tractor. The muffler was bent over to the left of the tractor, while the air intake also moved to the left, but was then moved to the vertical and extended up several feet, probably to get out of the "dust zone" created by the harvester.

International started to look at the postwar market for crawler tractors in 1944. A major restyling, converting the crawlers to the A series, was looked at; but the major news was a new, larger crawler tractor, the TD-24. Having the largest tractor in a category was always a source of pride and business for a company, particularly in the crawler field, where the allied equipment kept getting larger and larger. The TD-24 would give publicity value as well as profits for IH. Unfortunately, the execution of the project was flawed, as the crawler failed in early service with users.

Left
The T-40 was next fitted with an FBB six-cylinder engine. The tractor's sheet metal had to be extended forward, with an attractively sloped front hood. Again, the T-40 was undergoing the same changes that the experimental W-40s were receiving at the same time. Tractor Q-1742 weighed 10,340 pounds and was photographed August 20, 1931. *State Historical Society of Wisconsin*

The Model 15 TracTracTor makes its appearance on Valentine's Day 1930, one year to the day after a very famous and ugly event took place in Chicago. This tractor seems to have been an all-new from-the-ground-up program, rather than a variation of a wheel tractor. This tractor may have been part of the small tractor program that resulted later in the F-12/12 series, although the relationship is still unknown. After 25 T-15s were produced, the tractor was renamed the T-20. *State Historical Society of Wisconsin*

IH tried a six-cylinder FAB engine in the T-20. The track frame pivot on this tractor was also spring mounted, possibly indicating an interest in a higher-speed crawler. Tractor Q-2151 was photographed September 12, 1933. *State Historical Society of Wisconsin*

A TD-20? Well, why not. IH tried a 3 3/4x5 1/4 four-cylinder diesel engine in this tractor. There were also modifications to the equalizer on it. The diesel was probably a bit expensive at this point to put onto a small crawler. This tractor was photographed September 13, 1933. *State Historical Society of Wisconsin*

The TK-35 appeared suddenly in 1936. Here is one prototype, with a 3 3/4x4 1/2 six-cylinder kerosene engine. Visible are the clutch and brake compartment covers. One theory is that the T-35 series was derived from the smaller T-20 experiments with large engines, although this tractor has a much heavier duty undercarriage than the experimental T-20s of two years before. *State Historical Society of Wisconsin*

Here is a T-35 orchard TracTracTor. The operator's seat has been located on a much-extended PTO shaft, lowering the operator out of the way of branches. The controls have also been extended backwards as well. Photo taken October 11, 1937. *State Historical Society of Wisconsin*

The T-80 TracTracTor was one large piece of machinery for the time. The engine internal parts likely came from Waukesha. A split head of three cylinders apiece was used, as well as two air cleaners. This tractor may have also had a high- and low-range transmission—there's an awful amount of levers next to the seat! This photo was probably taken in January 1935. *State Historical Society of Wisconsin*

The T-65 looks like a junior version of the T-80, with some minor detail differences. Again, much of the engine for this tractor was probably supplied by Waukesha. Not much else is known of these interesting tractors. This T-65 was photographed February 5, 1935. *State Historical Society of Wisconsin*

Upper Right
Another, revised T-65. The air cleaners have been placed side-by-side instead of front and back, while the engine looks to have been heavily revised (the injection pump on the left side, instead of the right side of the engine). The undercarriage looks like it has also received some minor revisions. Tractor Q-2963 weighed 30,070 pounds and was photographed June 28, 1935. *State Historical Society of Wisconsin*

Lower Right
Next in the line of development was the TD-60 tractor. The tractor looks like a completely new tractor, with revised undercarriage, sheet metal, air cleaner, control equipment, and so on. Only the engine looks similar to the previous T-65. *State Historical Society of Wisconsin*

Another TD-60, with revised headlight position, a different air cleaner position, and some minor changes in the undercarriage. The decals have also been rearranged. This tractor was photographed May 19, 1937. *State Historical Society of Wisconsin*

As mentioned in the letter series tractor chapter (chapter 8), IH was experimenting with several new styling designs. Here is one of the first new styled crawlers, a TD-60. Even the undercarriage shows some signs of being restyled. New decal styles and hood trim accentuate the new design. This TD-60 was photographed December 9, 1937. *State Historical Society of Wisconsin*

A new crawler number appeared in early 1937. The TD-36 was probably the next development of the T-35 series. This one is fitted with the lighting attachment. It also has a wide-tread attachment, which doesn't show up in this photo, taken April 14, 1937. *State Historical Society of Wisconsin*

Upper Right
The TD-36 was not immune to the styling department. A somewhat different style is shown here, with "International" trim plates located on the driver's seat side, as well as the bottom of the radiator, where it probably would have either been hidden by attachments or quickly destroyed. Still, a neat looking tractor. This photo was taken January 24, 1938. *State Historical Society of Wisconsin*

Lower Right
Still another styling version of the T-36, this time with a decal "International" next to the driver's seat. This one also has the lighting attachment and wide tread. Tractor Q-3756 weighed 10,118 pounds and was photographed June 13, 1938. *State Historical Society of Wisconsin*

Yet another new number, the TD-25, was photographed June 21, 1938, and another styling theme makes an appearance. The removal of the side curtains at this stage probably was the result of the realization that they would have disappeared in the hands of users anyway. *State Historical Society of Wisconsin*

Finally, the Gas Power Engineering Department gets closer to the final product. Raymond Loewy's influence can definitely be seen in the simpler style of the tractor. The numbers of the tractor have also been rearranged into their final form. This tractor is apparently a production TD-14 with a new lighting system. This photo was taken October 18, 1938. *State Historical Society of Wisconsin*

Here is a TD-6 with a new orchard conversion. A slick, streamlined tractor that met a need, the TD-6 orchards are rare but sought-after tractors today. This one was photographed January 16, 1941. *State Historical Society of Wisconsin*

Above
This T-6 was modified to mount a pea harvester. The operator's seat and controls have been rotated 90 degrees, while the air cleaner has been extended to get out of the dust created by the harvester. This conversion was probably done for another manufacturer in the pea harvester business, or else a very large pea farmer who lived in a swamp! Seriously, peas have to be harvested when they're ready, wet or not. *State Historical Society of Wisconsin*

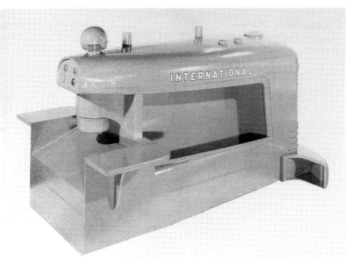

The A-series of IH crawlers was created in 1944 for the coming postwar market. Sheet metal, air cleaner, engine, and many other modifications were made to the basic tractors. Here is a wood mockup of some of the changes to the TD-18A, photographed on August 22, 1944. *State Historical Society of Wisconsin*

The TD-24 was another new model planned for the postwar market. This tractor was aimed to be just a bit bigger than the biggest Caterpillar tractor. Although it is not known when the first TD-24 TracTracTor was built, the first preproduction model was photographed October 9, 1945. *State Historical Society of Wisconsin*

This is another TD-24; a "pilot model," which is a term that IH didn't use much. Various changes can be seen, including a revised front bumper, undercarriage, and sheet metal. *State Historical Society of Wisconsin*

This is an extremely different TD-24, to say the least. Fitted with wheels instead of tracks, this tractor was produced for the Navy (or perhaps by the Navy!) and was probably built sometime in 1948 or 1949 (the Navy serial number for the beast was 48-03796). Nothing else is known about this extremely interesting tractor. *State Historical Society of Wisconsin*

This small model of a four-wheel-drive vehicle showed up in March 1952 at IH's Melrose Park Works. This may have been intended for Hough, but since Hough did most of its own engineering at this point, this vehicle may have been something else entirely. *State Historical Society of Wisconsin*

Left
Definitely in the running for the "cute" award, this T-2 crawler was almost certainly built with the 3x4 engine common to the Super A and C. Built at Melrose Park, this tractor was photographed February 24, 1950. Nothing else is known about the tractor. *State Historical Society of Wisconsin*

Here is a "4-Wheel Rubber Tired Tractor" at an unknown location in May 1954. This tractor was aimed at the military market, for use with an extremely interesting cargo trailer that connected to this tractor with a gooseneck. The trailer hitch was located between the operator's position and the engine on this tractor. *State Historical Society of Wisconsin*

Left
An unusual T-6 conversion, this Garret All Drive seems to have been intended for the logging industry. Not much else is known about the tractor, except that IH was very interested in it in the mid-1950s. This photograph was taken March 26, 1954. *State Historical Society of Wisconsin*

Appendix

Experimental and Preproduction Serial Numbers

While serial number information is scarce, some does survive. However, some information must be given about these tractors before you go on a treasure hunt. Because a tractor does not have an experimental serial number does not mean it was not an experimental tractor. IH used tractors off the line to test certain parts. In these cases, a Gas Power Engineering serial number tag (probably brass) will appear on the part itself. In addition, there may be instances of experimental tractors bearing high serial numbers of the tractor it was supposed to replace. It depended on how the tractor was produced. The Gas Power Engineering Department treated each major experiment as a separate job and assigned a serial number. A new air cleaner, for instance, may be Q-1032, while a new tractor prototype might be Q-1033. Gas Power Engineering used the prefix letter Q for all experimental jobs, whether cotton pickers, truck engines, refrigerators, or tractors.

Preproduction tractors were meant to test the tractor as it was supposed to be built. Sometimes the first series of preproduction tractors needed some changes, in which a second, and sometimes third series of preproduction tractors were brought out. These tractors were assigned serial numbers with the same letter prefix as the production tractor, but the numbers seem to fall into two categories—included in IH's production figures or not included in IH's production figures. Earlier tractors have the preproduction figures (and possibly experimental) included in the regular serial number series. For instance, the first 25 Mogul 8-16s are known to have been returned to the factory and scrapped, which was standard IH procedure for preproduction tractors starting at least by the early 1920s.

The scrapping policy was probably caused by farmers who owned certain tractors that IH never produced for regular sale. These farmers in some cases held on to them due to excellent performance, but occasionally had to have some new parts, which IH was obliged to supply years beyond when the experiments stopped. The four-wheel-drive version of the IH 8-16 is the known example of this happening. The farmer absolutely refused to sell the tractor back to IH for years. IH instituted the policy of trading in the preproduction or experimental tractors for tractors from the regular manufacturing run. In later years, IH would resume selling experimental tractors and preproduction tractors, but with the understanding that some parts might not be available, with the price reduced accordingly. In addition to these tractors, some of the older tractors escaped IH's efforts to find and scrap them due to their being sold and lost before IH came to find them. This accounts for the 1922 Farmall that survives.

Some of these are actually reasonable guesses: IH serial number records list starting serial numbers, but manufacturing authorizations often list production as starting at a higher serial number. The tractors between the official starting serial number and actual starting serial number are almost always preproduction tractors.

This list is far from complete. IH built numerous experimental tractors during production runs of the regular tractor. In addition, some tractors were sold to the public and then followed by the branch houses to test certain changes contemplated by IH, or to test parts made by a different vendor, as IH tried to have as many vendors as possible for outside-built parts. These parts included fan belts, spark plugs, wheels, and dozens of others.